A-LEVEL

STUDENT GUIDE

OCR

Geography

Earth's life support systems

Global connections

Peter Stiff and David Barker

HODDER
EDUCATION
AN HACHETTE UK COMPANY

Hodder Education, an Hachette UK company, Blenheim Court, George Street, Banbury, Oxfordshire OX16 5BH

Orders

Hachette UK Distribution, Hely Hutchinson Centre, Milton Road, Didcot, Oxfordshire, OX11 7HH

tel: 01235 827827

e-mail: education@hachette.co.uk

Lines are open 9.00 a.m.–5.00 p.m., Monday to Friday. You can also order through the Hodder Education website: www.hoddereducation.co.uk

ISBN 978-1-4718-6401-8

First printed 2016

Impression number 5

Year 2022

Cover photo: Richard Carey/Fotolia

Typeset by Integra Software Services Pvt Ltd, Pondicherry, India

Printed and bound by CPI Group UK)Ltd, Croydon, CR0 4YY

Hachette UK's policy is to use papers that are natural, renewable and recyclable products and made from wood grown in well-managed forests and other controlled sources. The logging and manufacturing processes are expected to conform to the environmental regulations of the country of origin.

Contents

Content Guidance

Questions & Answers

Getting the most from this book

Exam-style questions

Commentary on the questions

Tips on what you need to do to gain full marks, indicated by the icon ⓔ

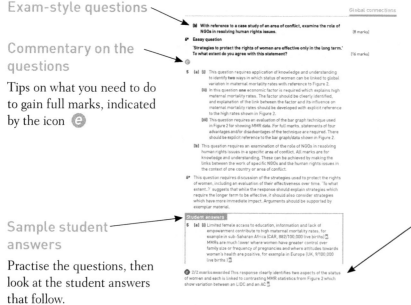

Sample student answers

Practise the questions, then look at the student answers that follow.

Commentary on sample student answers

Read the comments (preceded by the icon ⓔ) showing how many marks each answer would be awarded in the exam and exactly where marks are gained or lost.

■ About this book

Much of the knowledge and understanding needed for AS and A-level geography builds on what you have learned for GCSE geography, but with an added focus on geographical skills and techniques, and concepts. This guide offers advice for the effective revision of A-level topics **Earth's life support systems** and **Global connections** (Trade in the contemporary world, Global migration, Human rights, and Power and borders).

The external exam papers test your knowledge and application of these aspects of physical and human geography. More information on this is given in the Questions & Answers section at the back of this book. To be successful in these topics you have to understand:

■ the key ideas of the content
■ the nature of the assessment material — by reviewing and practising sample structured questions
■ how to achieve a high level of performance within them.

This guide has two sections:

Content Guidance — this summarises some of the key information that you need to know to be able to answer the examination questions with a high degree of accuracy and depth. In particular, the meaning of key terms is made clear and some attention is paid to providing details of case study material to help to meet the spatial context requirement within the specification. Students will also benefit from noting the **Exam tips** that will provide further help in determining how to learn key aspects of the course. **Knowledge check** questions are designed to help learners to check their depth of knowledge – why not get someone else to ask you these?

Questions & Answers — this includes some sample questions similar in style to those you might expect in the exam. There are some sample student responses to these questions as well as detailed analysis, which will give further guidance in relation to what exam markers are looking for to award top marks.

The best way to use this book is to read through the relevant topic area first before practising the questions. Only refer to the answers and examiner comments after you have attempted the questions.

Content Guidance

This section outlines the following areas of the OCR A-level geography specification:

- Earth's life support systems
- Global connections
    - Trade in the contemporary world
    - Global migration
    - Human rights
    - Power and borders

■ Earth's life support systems

How important are water and carbon to life on Earth?

Water and carbon support life on Earth and move between land, oceans and atmosphere.

The importance of water

Comparisons between Earth and the other planets in our solar system soon show that the widespread availability of water is a distinguishing feature of our planet. Due to its distance from the Sun, the amount of energy intercepted by the Earth allows liquid water to exist in vast quantities. Water vapour and solid water (ice) also exist and transfers between these three stores of water occur. Liquid water allows organic molecules to mix which can then form more complex structures which are vital in the evolution of life.

The biosphere relies on liquid water to be available partly because all living organisms need it but also because of interactions between water and the atmosphere.

Water helps regulate temperatures. Oceans (71% of Earth's surface) absorb, store, transport and release heat energy. In the atmosphere, clouds reflect about 20% of incoming short-wave radiation back out to space. Water vapour, a greenhouse gas, absorbs outgoing long-wave radiation thereby helping create average global temperatures almost 15°C higher than they would be otherwise. Life as we know it would not survive at lower temperatures.

The uses of water for flora, fauna and people

Water is the most abundant molecule in living cells. Between 65–95% of all living organisms (flora – plants, fauna – animals and people) is water. Photosynthesis, the fundamental process that virtually all life relies on, requires water.

Water carries substances in and out of living cells, circulating energy and waste products for example. It helps many biochemical reactions to take place such as **hydrolysis** (addition of water) and **condensation** (removal of water) which break

> The biosphere is the thin zone in which life occurs. It extends only a few hundred metres above the surface, penetrates not very far below ground but deeper into the oceans.

> Photosynthesis is the process whereby plants combine carbon dioxide (CO_2) from the air, water and minerals and using sunlight make complex molecules such as glucose.

down or join up molecules in cells. Flora require water to help support their tissues and without sufficient water, plants wilt. Plants transpire water, which is a vital process moving water around the biosphere.

Fauna and people also depend on water in maintaining their right internal conditions. Sweating for example helps regulate internal temperature.

Water pays an essential role in economic activities such as power generation, many manufacturing processes, agriculture (see Table 1), sewage disposal and many leisure pursuits.

Table 1 Water required in production of selected agricultural products

Product	Litres of water required to produce 1 kg
Chocolate	17,000
Beef	15,500
Cheese	3,200
Potatoes	300
Egg	200 (per egg)

To produce 1 tonne of steel takes between 95,000 and 150,000 litres of water. Manufacturing a cotton shirt or blouse requires up to 2,700 litres of water. Much of the water used can be recycled although it can carry pollutants. Water is as important as energy to human society.

The importance of carbon

Carbon is one of the most abundant chemical elements on Earth. Its importance lies in its ability to bond with many other elements and it is estimated that it forms the basis of 95% of all known compounds. Carbon is ubiquitous on Earth, being found in carbonate rocks such as limestones, in sea bed sediments, dissolved CO_2 in oceans, as CO_2 gas in the atmosphere and throughout the biosphere.

Carbon is the building block for life

Life on Earth as we know it is based on carbon. It is present in nearly every molecule and structure within living organisms in the natural world and in humans.

Carbon is the foundation of virtually all human activities. Currently, some 80% of the world's primary energy is supplied by fossil fuels, such as coal, oil and natural gas. Fuelwood is an important energy source for millions of people, mainly in low-income developing countries (LIDCs).

Carbon is a significant raw material for manufacturing. Oil is fundamental in the production of plastics, paints and synthetic fibres such as polyester. Timber is used in construction and paper-making while agricultural crops not only provide food but also materials such as cotton and oils for products such as soap.

Water and carbon systems

Water and carbon are cycled through open and closed systems. Systems consist of inputs, stores and processes within the system, and outputs. These components are linked together by flows of energy and materials through the system. Open systems

Transpiration is the movement of water vapour molecules out of plants through 'pores' known as stomata. It maintains a plant's internal condition as regards water content and helps cool the plant.

Knowledge check 1

State **three** ways water is important for life on Earth.

Ubiquitous means being present, found or appearing everywhere.

are those where energy and materials move in and out of the system. In a closed system, only energy can transfer across its boundary.

At the global scale, both the water and carbon cycles are closed systems. The Sun's energy flows across their boundaries from outside of the Earth to provide energy to 'drive' the cycles. In smaller-scale systems such as a drainage basin or a forest, both energy and materials move in and out. Precipitation falls into the drainage basin and water flows out, often into the sea. Fauna move in and out of a forest as does water.

The water and carbon cycles are systems with inputs, stores, processes and outputs.

The global water cycle

Stores

The greatest volume of water is stored in the oceans, some 97% of all water on Earth. Two percent of water exists as snow and ice (ice-caps, ice sheets and glaciers such as on Antarctica and Greenland). Water stored in a frozen (solid) state, part of the **cryosphere**, accounts for 75% of all the **fresh** water on Earth. Groundwater (**aquifers**) accounts for 0.7% of all water and 20% of fresh water (Figure 1).

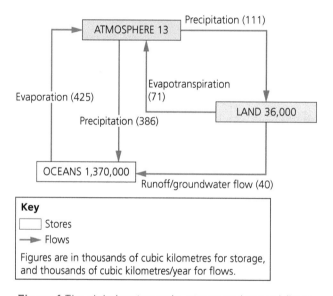

Figure 1 The global water cycle: stores and annual flows

All the surface water in lakes and rivers only accounts for just over 0.01% of all water, with the remaining water stored in soils, flora, fauna and humans and the atmosphere. While the proportion of water stored in the atmosphere at any one time is negligible, there is such a rapid turnover of water in the atmosphere that it is a very important factor in the water cycle.

Inputs and outputs

Just over 500,000 km³ of water circulates between the principal water stores due to some key processes.

- **Evaporation** and **transpiration**: water evaporates from surface water (oceans, lakes, rivers) and from surfaces on land, natural and human-made. Plants transpire

> **Exam tip**
>
> Make sure you understand the difference between open and closed systems. It is helpful if you can give examples of inputs, stores and processes and outputs.

> The **cryosphere** is any area where water exists as snow and/or ice. This can be on land, frozen lakes and rivers and sea ice.

> **Aquifers** are rocks, such as chalk and some sandstones that can store and transmit water.

water through their leaves. The term **evapotranspiration** is used to describe these two processes transferring water into the atmosphere.

- **Condensation, precipitation, ablation** and **sublimation**: water vapour in the atmosphere becomes liquid water due to condensation. Water can then be transferred from the atmosphere to the surface as precipitation. Ablation is the process by which water is lost from snow or ice, such as melting, calving of icebergs or evaporation. Sublimation is when water changes from its solid state (ice) to its gaseous state (water vapour).
- **Runoff** and **infiltration**: water on the surface drains across it as runoff. Runoff collects into streams and rivers with most rivers eventually flowing into the oceans. Most precipitation ends up as runoff after infiltrating into, and flowing through, the soil.
- **Percolation**: the transfer of water into aquifers where it is known as groundwater. Eventually this water can emerge as runoff from springs.

Knowledge check 2

Why is atmospheric water such an important element in the water cycle despite it being such a relatively small store?

The global carbon cycle

The land and sea stores (sinks) dominate the carbon cycle (Table 2). Virtually all carbon (99.9%) is stored in sedimentary rocks such as limestones. The carbon present in the atmosphere, oceans, soil and biosphere circulates relatively rapidly (Figure 2).

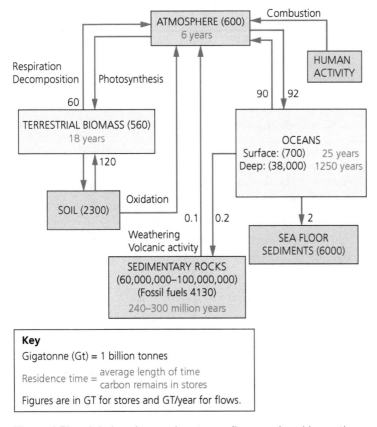

Key

Gigatonne (Gt) = 1 billion tonnes

Residence time = $\dfrac{\text{average length of time}}{\text{carbon remains in stores}}$

Figures are in GT for stores and GT/year for flows.

Figure 2 The global carbon cycle: stores, flows and residence times

Table 2 Carbon stores (sinks)

Store	Carbon in store (billion tonnes)
Sedimentary (carbonate) rocks and deep ocean sediments	100,000,000
Oceans (dissolved CO_2)	39,000
Fossil fuels (coal, oil and natural gas)	4,000
Soils and peat	1,500
Atmosphere (CO_2)	720
Land plants (e.g. forests and grasslands)	560

There are two strands to the carbon cycle: one slow, the other fast.

The slow carbon cycle

Atmospheric CO_2 is absorbed by the oceans. Marine organisms such as coral, shellfish and phytoplankton combine carbon with calcium so that they can build their shells and skeletons from calcium carbonate. When these organisms die, they sink to the ocean floor and accumulate. Over millions of years, heat and pressure convert these deposits to carbon-rich sedimentary rock.

Typically, carbon is stored in rocks for some 150 million years. Some rock is subducted and melted at plate boundaries and carbon is vented out to the atmosphere as CO_2 via volcanic eruptions. Rocks at or near the surface are chemically weathered which can release CO_2 into the atmosphere. Carbon can also be carried away in dissolved forms by streams and rivers, eventually ending up in the oceans.

Carbon can accumulate in sinks such as fossil fuels. Coal, lignite, oil and natural gas are sinks which store carbon for millions of years.

The fast carbon cycle

Although carbon storage in the atmosphere, plants, soil and peat is relatively small, these stores are crucial in the carbon cycle.

Transfers of carbon between the atmosphere, oceans, living organisms and soils are some ten to one thousand times faster than those in the slow carbon cycle. The key components are land plants and phytoplankton in the oceans.

Phytoplankton absorb CO_2 from the oceans by photosynthesis after which it can be transformed into carbohydrates and stored in their tissues. Respiration by plants and animals as well as decomposition of dead organic material returns CO_2 to the atmosphere.

The oceans and atmosphere exchange carbon, with CO_2 dissolving in surface water and a return flow of CO_2 occurring via evaporation.

Processes and pathways of the water and carbon cycles

The water and carbon cycles have distinctive processes and pathways that link the components of the systems.

Knowledge check 3

Why is the burning of fossil fuels such a concern in terms of carbon cycling?

Peat is partly decomposed organic matter that has accumulated in waterlogged and therefore anaerobic conditions.

Phytoplankton are tiny, sometimes microscopic, plant organisms that float and drift in the oceans, capturing the Sun's energy through photosynthesis.

Exam tip

It is important that you are able to outline the main differences between the slow and fast carbon cycles. Always give details such as the names of the different stores (sinks).

Processes of the water cycle

The water balance

Within a drainage basin the flows of water can be summarised in the following equation.

Precipitation (P) = Evapotranspiration (E) + Streamflow (Q) +/- Storage

Flows of water

Evaporation

This is when liquid water changes into water vapour and is a process requiring considerable energy. The energy does not heat the water but transfers into latent heat and is released when condensation occurs. Vast amounts of heat energy are carried into the atmosphere from tropical oceans. The warm air blows away from the tropics to the north and south where cooling releases latent heat. In this way mid- and high latitudes receive energy which helps make them less cold than they would be otherwise. Evaporation is also vital in transferring water from oceans to land.

Transpiration

This is the diffusion of water vapour from a plant to the atmosphere, mostly via pores in leaves known as stomata. It contributes about 10% of atmospheric moisture. Air temperature, wind speed and how much water is available to the plant, are critical controlling factors. It can vary considerably from one season to another, for example between summer and winter in a deciduous forest in the mid-latitudes.

Condensation

This is when water vapour changes to liquid water. When water vapour in the atmosphere cools to dew-point, water droplets and clouds appear. At or near ground level, condensation results in mist, fog and dew.

Cloud formation

Cooling of air and its water content are important factors in cloud formation. Cooling can occur by the following processes.

- Convection: air is heated at the surface, rises in columns and cools. This tends to produce cumulus clouds.
- Frontal uplift: warm air rises along a front, either a warm or cold front, a distance of several hundreds of kilometres. Stratus clouds can be formed at lower levels with cirrus clouds at high altitudes.
- Advection: a large mass of air moves horizontally over a cooler surface, often the ocean. The atmosphere is thereby cooled below the dew-point resulting in stratus clouds forming.
- Orographic uplift: when air is forced to rise over a range of hills or mountains. If sufficient cooling occurs, condensation takes place.

Dew-point is the temperature at which air becomes fully saturated. Further cooling results in condensation.

Cumulus clouds have flat bases and extend vertically up through the atmosphere. Some can be 10–15 km from top to bottom.

Stratus clouds exist as a widespread dense horizontal layer.

Cirrus clouds are made up of ice crystals and have a wispy appearance.

Cloud formation and lapse rates

A lapse rate is the change in temperature with increasing height through the lower part of the atmosphere. There are three types of lapse rate (Table 3).

Table 3 Types of lapse rate

Environmental lapse rate (ELR)	The actual change in temperature at any given place or time. It is what would be measured by a thermometer carried upwards by a balloon. Approximately 6.5°C / km.
Dry adiabatic lapse rate (DALR)	The change in temperature of a parcel of dry air rising up through the atmosphere when no condensation occurs. Approximately 10°C / km.
Saturated adiabatic lapse rate (SALR)	The change in temperature of a parcel of air rising up through the atmosphere when condensation is occurring. Approximately 7°C / km, a lower rate than the DALR because condensation releases latent heat.

Adiabatic refers to the situation when no heat energy is transferred into or out of the rising parcel of air. As the volume of the parcel increases with height, the same amount of energy has to 'heat' more air, therefore the temperature of the parcel of air decreases.

As long as a column or parcel of air is warmer than the surrounding air, it will continue to rise and cool. When the temperatures are the same, no further uplift is possible. A parcel of air colder than its surroundings will fall until it has warmed to the same temperature as its surroundings.

Just because a cloud exists does not mean that precipitation will fall. Further cooling and instability are required for water to flow from the atmosphere store to the surface.

Precipitation

Precipitation is any water that transfers from the atmosphere to the land or oceans, such as rain, snow, hail, sleet or drizzle.

In mountainous areas and high latitude regions (e.g. the Arctic) precipitation often falls as snow. Frozen water is then stored on the surface, sometimes for several months, before it melts and flows through the drainage basin.

The intensity with which precipitation falls affects how fast water flows through a drainage basin. High intensity rain (e.g. > 10–15 mm/hour) flows rapidly over the surface into streams and rivers. This is common in tropical rainforests.

How long precipitation falls for (duration) influences the speed of water flow. Rain that persists for many hours can lead to fast flows of water whereas rain falling for ten minutes will tend to be absorbed by the soil and plants relatively quickly.

In some regions, such as the Sahel or much of the Indian sub-continent, the climate is divided into a wet and a dry season. During the wet season, precipitation tends to be intense and long lasting leading to rapid flows of water and river flooding. The dry season can result in surface flows drying up completely and storage in the soil becoming exhausted.

Exam tip

Being able to mention different types of precipitation and allocate these to actual locations will add authority to your discussions about the water cycle.

Interception

The temporary storage of precipitation on vegetation or buildings before water reaches the surface is called interception. Some of this water evaporates straight back into the atmosphere. Interception by vegetation varies considerably depending on the type of vegetation and season. Beech trees intercept just over twice as much precipitation in summer than in winter, about 40% compared to about 20%. Pine trees intercept more than beech trees no matter what the season.

Buildings are designed to intercept any water that falls on their roofs. Water is then rapidly transferred to underground pipes and channelled away via drains to streams and rivers.

Infiltration, throughflow, percolation, groundwater flow and runoff

Infiltration is the process of water soaking into the soil surface. Once in the soil, water moves relatively slowly compared to when it is flowing over the surface. Different types of soil have different infiltration capacities. Clay soils, for example, absorb water at slower rates than do sandy soils.

Once in the soil, gravity forces water to move downhill as throughflow towards stream and river channels. Some water also moves deeper into underlying rocks by percolation. As in the soil, gravity moves groundwater flow through rocks towards channels.

Different rock types vary in their capacities to absorb and store water (Figure 3).

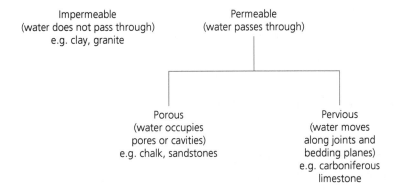

Figure 3 Rock type and water movement

Soil and groundwater stores and flows tend to vary. **Water tables** rise during periods of high rainfall and low evapotranspiration and vice versa.

All water that drains from an area is known as runoff. This process can be considered at any scale, from the flow of continental rivers such as the Ganges, through smaller rivers such as the Exe down to a local stream. Surface runoff can be seen occurring from a sports field or playground.

Water flow occurs underground via throughflow and groundwater flow. It also happens across the surface. When the soil store is full, any additional input of water results in saturated overland flow. Overland flow occurs when rainfall is so intense that more water falls onto the surface than the soil has capacity to absorb. For

Knowledge check 4

Why do interception rates for vegetation vary with the season?

The **water table** is the boundary between saturated and unsaturated conditions underground.

example, semi-arid areas experience short but torrential storms which the baked-hard surface cannot absorb. Temporary streams and rivers flow as a result.

Ablation

This is the loss of water stored as ice or snow through melting, runoff, evaporation, sublimation, calving of icebergs or the removal of loose snow by wind.

In mid-latitude regions, melting is the most important process. In the Antarctic, iceberg calving is the most significant form of ablation.

Calving occurs when a glacier reaches the sea and ice masses break off and float away.

Processes of the carbon cycle

Carbon flows between stores, such as the atmosphere and rocks, are usually referred to as exchanges or fluxes.

Precipitation

Carbon dioxide in the atmosphere can be dissolved in rainwater to form weak carbonic acid. Carbon flows back to the oceans and land when precipitation happens. This entirely natural process is being affected by the rising concentrations of CO_2 in the atmosphere due to anthropogenic emissions which result in increases in rain's acidity. This in turn is partly responsible for increases in the acidity of ocean surface waters.

Anthropogenic means caused by human activities.

Photosynthesis

Land plants and marine phytoplankton fix atmospheric CO_2 converting light energy from the Sun into chemical energy via photosynthesis.

carbon dioxide + water ⟶ glucose + oxygen

$$6CO_2 + H_2O \longrightarrow C_6H_{12}O_6 + 6O_2$$

Glucose is used in several ways:
- as a building block of cell walls in the form of cellulose, an important carbon store
- for respiration inside plant cells to provide energy
- to be converted into fats and oils which act as energy storage
- to be converted into amino acids used to synthesise proteins.

Respiration

The process releases energy by breaking down complex molecules such as sugars and CO_2 is given off back into the atmosphere. It is the reverse of photosynthesis.

glucose + oxygen ⟶ carbon dioxide + water

$$C_6H_{12}O_6 + 6O_2 \longrightarrow 6CO_2 + H_2O$$

Together, photosynthesis and respiration are vital aspects of the fast carbon cycle, exchanging a one thousand times greater volume of carbon than is moving through the slow carbon cycle.

Exam tip

You do not have to learn the chemical formulae but you must be able to describe the processes accurately using words.

Decomposition

When living organisms die, their cells begin to break down. Processes of decomposition are carried out by microorganisms (mainly bacteria and fungi) and carbon is released as CO_2. Rates of decomposition vary considerably around the world. Humid locations where temperatures are warm all year round, such as tropical rainforests, have fast rates whereas in the Arctic tundra decomposition is slow. Arid regions also have slow rates due to the lack of water for chemical reactions.

Combustion

Combustion occurs when organic material reacts or burns in the presence of oxygen. Carbon is released as CO_2.

In many locations, combustion occurs naturally by lightning strikes starting wildfires and is a key process in maintaining the health of ecosystems. Organic debris that has accumulated is burned which results in increased carbon and nutrient cycling. New habitats are created and fresh growth stimulated. The coniferous forests of the Rocky Mountains and the grassland plains in east Africa are two ecosystems in which natural combustion is of great benefit.

Humans use combustible materials derived from plants to generate power. Use is made of recently alive plants such as wood from trees or farmed energy crops. Biomass fuel is regarded as carbon neutral as, although carbon is released into the atmosphere during combustion, when combustion occurs the next crop of plants has already been planted and is absorbing CO_2 from the atmosphere. No net increase in atmospheric CO_2 occurs providing the system is sustainably managed.

The stark contrast with the combustion of fossil fuels is in the time scales involved. While fossil fuels are plant based, they represent a carbon sink millions of years old. When combusted, oil, coal and natural gas release this stored carbon leading to a net increase in CO_2. Some 10 GT of CO_2 flows from the long-term geological store to the atmosphere, oceans and biosphere annually.

Weathering

This is the in situ breakdown of rocks at or near the Earth's surface by physical, chemical and biological processes. Most chemical weathering processes involve rainwater which is weak carbonic acid (H_2CO_3). The CO_2 is dissolved from the atmosphere and biogenic CO_2 is also absorbed as water trickles through the soil making the carbonic acid stronger.

Carbonation or solution weathering is the principal chemical weathering process affecting carbonate rocks such as the limestones, for example chalk and carboniferous limestone. These rocks are mainly composed of calcium carbonate which is insoluble. Once it has reacted with carbonic acid, it changes into calcium bicarbonate which is soluble and can therefore be dissolved and carried away in running water.

calcium carbonate + carbonic acid $\longrightarrow$ calcium bicarbonate

$$CaCO_3 + H_2CO_3 \longrightarrow Ca(HCO_3)_2$$

Exam tip

It is important to appreciate the ways in which natural combustion can benefit an ecosystem.

Biomass is biological material derived from living or recently living organisms. In the context of power generation it is usually plant material.

Knowledge check 5

What is the significance of the contrast between human use of biomass and fossil fuels?

Biogenic CO_2 is produced by the respiration and decomposition of organisms in the soil.

Some 0.3 billion tonnes of carbon are transferred from rocks to the atmosphere and oceans each year by chemical weathering.

Physical weathering involves no chemical changes to rocks. However, it results in smaller fragments of rock being broken off which then have a greater surface area to be attacked by chemical processes. Physical weathering also makes and widens cracks in rocks, for example by frost action, thereby allowing more water to access the rock. This allows chemical weathering to be more effective.

Biological weathering is the result of the action of organic acids on rocks. **Chelation** is a key process in which humic acids attack minerals in rocks, leading to rock breakdown. This process is especially active in tropical rainforests due to warm, humid conditions and plenty of decaying organic matter to produce humic acid.

Natural storage of carbon

Carbon sequestration takes place naturally and in a variety of ways.

- Vegetation: photosynthesis fixes carbon from the atmosphere in plants. Tropical rainforests and high latitude coniferous forests, as well as extensive areas of grassland, such as the steppes, prairies or pampas, represent vast carbon sinks.
- Sediments: weathered rock containing carbon accumulates at the bottom of the deep oceans and in lakes. Here these sediments will, over geological time, form new sedimentary rock.

The oceans play a particularly important role in carbon sequestration in two ways.

1 **Physical (inorganic) pump**

   Carbon dioxide is absorbed by the surface waters. Ocean currents transfer this water towards the poles where it cools. The water, containing the dissolved CO_2, becomes denser and sinks. This downwelling of water occurs in only a few locations, one of which is in the North Atlantic between Greenland and Iceland. The carbon can remain at great depth for centuries. Eventually the bottom water is carried upwards and when it reaches the surface, CO_2 diffuses back into the atmosphere, for example off the coasts of Peru and Chile.

2 **Biological (organic) pump**

   On a global scale, nearly half of all the carbon fixed by photosynthesis occurs in the oceans. Phytoplankton which float close to the surface perform the same function of carbon capture as do land plants. This carbon then passes through the marine ecosystems and much of it accumulates on the ocean floor when organisms die.

   Marine organisms that make shells do so using carbonate minerals from sea water. When these creatures die, their shells accumulate on the ocean bed where they can eventually form new carbonaceous rocks.

Sequestration is the capture and long-term storage of carbon from the atmosphere. It occurs naturally but humans are trying to find ways of achieving this in order to reduce levels of atmospheric CO_2.

Knowledge check 6

What is the significance of the oceans in carbon sequestration?

How do the water and carbon cycles operate in contrasting locations?

It is important to identify the physical and human factors that affect the water and carbon cycles in two contrasting locations.

Case studies: tropical rainforest and Arctic tundra

For this part of the specification you are required to have two case studies. One is of a tropical rainforest area and the other an Arctic tundra area.

Your case studies must illustrate:
- the specific water and carbon cycles operating in each location
- the range of factors that influence the water and carbon cycles
- the role of human activities operating in each location.

You should be able to describe and explain for each location:
- the rates of flow of water and carbon and the distinct stores of water and carbon
- the physical factors affecting the flows and stores including temperature, rock permeability and porosity, relief, vegetation, organic matter in the soil, and the mineral composition of rocks.

You should be able to describe and explain for the tropical rainforest location:
- in the context of one drainage basin, changes to the flows and stores within the water cycle caused by natural and human factors such as deforestation and farming
- the impacts of human activities such as deforestation and farming on flows of carbon and stores of carbon in the soil and nutrient stores
- strategies to manage the location that have positive effects on the water and carbon cycles such as afforestation and various agricultural techniques, for example rotational cropping and greater integration of livestock and arable operations.

You should be able to describe and explain for the Arctic tundra location:
- seasonal changes in the water and carbon cycles
- the impacts of the developing oil and gas industries on the water and carbon cycles
- strategies to manage and moderate impacts from the oil and gas industries.

Examples of places suitable for case studies are:
- rainforest regions in central and south America (River Amazon), central Africa (River Zaire) and southeast Asia (Mekong River)
- Arctic tundra such as northern Canada, Alaska, northern Scandinavia, northern Siberia.

The **Arctic tundra** is a vast, relatively level, treeless and marshy region, usually with permanently frozen subsoil. It extends from where the coniferous forests end to the edge of the Arctic Ocean.

Exam tip

As well as the facts and figures of your case studies, it is important that you are able to use technical terms, such as 'permafrost' and 'leaching', with authority.

Change over time in the water and carbon cycles

How much change occurs over time in the water and carbon cycles? Human factors can disturb and enhance the natural processes and stores in the water and carbon cycles.

Dynamic equilibrium

Natural systems are constantly changing. Energy flows through them and materials do not stay in the same state. At a variety of scales both the water and carbon cycles exist in dynamic equilibrium as long as they are unaffected by human activities.

In a drainage basin, an increase in water input — a period of intense rain for example — will lead to more water entering the stores and if these overflow, then river levels rise and flooding may result. Eventually, the rain stops, the rivers discharge declining volumes of water, floods recede and the drainage basin system reverts back to its state before the storm.

At the global scale, some increase in atmospheric CO_2 can increase rates of photosynthesis. Plants therefore grow vigorously and carbon is stored in their tissues. This feedback decreases the quantity of CO_2 in the atmosphere and a state of dynamic equilibrium is achieved.

The effects of land-use changes

Land-use changes affect the water and carbon cycles.

Urbanisation

When an area of farmland, woodland or marsh, for example, is built over, stores and flows of water are significantly altered. Most artificial surfaces such as tarmac or concrete are impermeable and designed to shed water as quickly as possible. Little or no water can enter either the soil or groundwater stores. Urban areas have a network of gutters, drains and sewers designed to move water rapidly to a stream or river. The result of an input of rain is that stream and river discharge levels rise very quickly, sometimes leading to flooding.

Much building development is attracted to floodplains. These flat areas allow relatively straightforward construction which keeps costs down. However, the water storage capacity of the floodplain is severely reduced as permeable surfaces are replaced by impermeable ones.

The removal of vegetation and the covering over of soil by buildings has severe impacts on carbon cycling. Very little carbon is sequestered in urban areas as natural stores such as vegetation and soils are much reduced. The actual manufacture of building materials and construction tends to involve the combustion of large quantities of fossil fuels. Urban areas and the activities operating in them such as transport and space heating also tend to involve the release of CO_2 in the atmosphere.

Farming

When a natural ecosystem is altered by the introduction of farming, both water and carbon cycles are affected. Water can be abstracted from surface and underground stores for use in irrigation. Most of this water flows as evapotranspiration into the atmosphere or drains through the soil.

Rainfall interception and evapotranspiration by crops is less than in forest or grassland ecosystems. Ploughing increases evaporation, resulting in soils drying out. If furrows are ploughed following a field's slope rather than across the gradient, runoff and soil erosion can occur at increased rates. Water flows quickly off fields

Dynamic equilibrium suggests that a system can adapt to changes, for example in the inputs or stores, but that through feedback a balance or equilibrium is achieved.

Exam tip

It is not enough simply to state that dynamic equilibrium exists. You need to be able to give worked examples in both water and carbon cycles.

where underground drainage systems are installed. Large-scale and heavy agricultural machinery can compact soils reducing infiltration rates. All these factors can increase the peak flows in streams and rivers and contribute to flooding.

Deforestation reduces carbon storage as biomass, both above and below ground. When soils are exposed, oxidation of organic matter is accelerated so that more CO_2 is released into the atmosphere. The harvesting of crops removes carbon so that only a relatively small quantity of carbon accumulates in the soil as **humus**.

The impact of farming on previously grassland ecosystems is less severe in terms of carbon cycling. Depending on how sustainable the type of farming is there still is less carbon stored and a net loss of carbon.

Forestry

Interception rates are higher in managed plantations than in natural woodland, because planting is denser. Most plantations in the UK are evergreen conifer species such as Sitka spruce whose leaves are in the form of needles. These factors also lead to increased interception. In turn water stored on the tree surfaces such as the needles, flows back to the atmosphere through evaporation. Transpiration rates also tend to be higher from plantations than the upland moorland they replace.

Less water reaches channels from forestry plantations and that which does eventually flow as streams and rivers takes a long time travelling through the drainage basin. But where clear felling is practised, evapotranspiration is greatly reduced and runoff can be very high.

Forests are capable of sequestering carbon for centuries both in the trees and the woodland soil. In a typical UK plantation, mature trees contain an average of 170–200 tC/ha^{-1} which is ten times higher than grassland. Soils under forests can store even greater quantities of carbon, as much as 500 tC/ha^{-1}.

Trees do not maintain the same rate of carbon capture through their lives. After about 100 years the quantity of carbon stored in the tree levels off and is balanced by the carbon in the leaves and twigs falling off the tree decomposing on the forest floor. Commercial forestry often harvests 100-year-old trees and then replants, thus continuing the cycle. Well-managed woodland tends to have healthier and faster growing trees than natural forests and so carbon capture is at a higher rate.

The impact of water extraction

How does water extraction impact on flows and stores in the water cycle? Whatever human activity you think of, somewhere along the line, water is consumed. There are two basic impacts of water extraction on flows and stores in the water cycle (Table 4).

Table 4 Impacts of water extraction on flows and stores in the water cycle

Type of extraction	Impacts
Surface	Reduces channel flows, increases sediment deposited, dissolved O_2 levels fall and water temperatures rise threatening organisms, pollutants are less diluted, fewer floods reduce the supply of alluvium to floodplain
Groundwater	Reduces level of water table in aquifers and artesian basins, springs and streams dry up, ecosystems are threatened, can extract 'fossil' water, i.e. water stored underground, for hundreds of years faster than it is replenished

Humus is the organic matter in the soil derived from the breakdown of organic matter.

An **artesian basin** is an aquifer held in a syncline or basin-like geological structure.

Where and when the extraction of water is at a rate faster than the rate of recharge, water shortages result. In some semi-arid regions this is due to population growth and rising demand. In north Africa, aquifers deep underground are being exploited with water extracted that has been stored for hundreds of years.

The impact of fossil fuel combustion and carbon sequestration

Fossil fuel combustion and carbon sequestration has an impact on flows and stores of carbon.

Fossil fuel combustion

Energy is needed to produce all goods and services. Fossil fuels have dominated energy supply for the past two centuries. Some 85% of global energy consumption currently comes from coal, oil and natural gas despite developments in nuclear power and renewables such as wind and hydro-electric power (HEP). The world's population will increase to about 9 billion by 2050 and many millions will experience rising standards of living which consume more energy.

Combustion of fossil fuels contributes 10 billion tonnes of CO_2 to the atmosphere each year. Since 1750, anthropogenic CO_2 has raised concentrations of the gas from 280 parts per million (ppm) to 400 ppm. This is despite vast amounts of the released carbon being absorbed in the oceans or taken up by living organisms such as trees (Figure 4).

Where our carbon emissions have come from: carbon emission sources 1750–2012 (Gt CO_2)

Where our carbon emissions have gone: carbon emission sinks 1750–2012 (Gt CO_2)

Figure 4 Carbon emissions and sinks 1750–2012

Anthropogenic CO_2 emissions may be less than 10% of the natural flow of carbon from the biosphere and oceans into the atmosphere, nevertheless they have a very significant impact. For example, CO_2 molecules add greatly to the trapping of outgoing long-wave radiation and so enhance the natural greenhouse effect of the atmosphere. In turn this leads to global warming.

Carbon sequestration

Carbon capture and storage (CCS) aims to reduce CO_2 emissions from fossil fuel combustion, in particular in gas- and coal-fired power stations. It is also a possible solution to the carbon flows from heavy industries such as iron and steel and cement manufacturing and oil refining.

The CSS process has three parts: separating CO_2 from waste gas emissions, transporting the captured CO_2, and secure long-term storage of CO_2 underground. Technologies to achieve CSS exist but obstacles also exist, such as development costs and the political will to make it happen. In November 2015 the UK government cancelled their financial support of such projects. There are, however, some 15 large-scale CSS projects around the world. In Canada CSS captures carbon from coal combustion and oil extraction from tar sands, and in Norway from oil refining.

Exam tip

Keep up to date with projects such as CSS by regular research using websites such as those maintained by the International Energy Agency and the CSS Association.

Feedback within and between the water and carbon cycles

There are positive and negative feedback loops within and between the water and carbon cycles. Change in any system brings about either positive or negative **feedback** responses:

- positive feedback occurs when a change (e.g. to an input or a store) causes further changes in the system which can make the system unstable
- negative feedback occurs when a change causes the system to alter so that equilibrium (balance) is recovered and the system does not change.

The carbon and water cycles operate at a variety of scales, so the feedback within them also takes place at different scales.

Feedback is the automatic response of a system when a change to the flow of energy or materials through it takes place.

Feedback in the water cycle

Globally increasing temperatures result in higher rates of evaporation. Water vapour is a greenhouse gas therefore more long-wave radiation is absorbed and the atmosphere warms. Higher temperatures lead to more evaporation and so on. This is an example of positive feedback.

Alternatively, negative feedback can occur from more water vapour in the atmosphere. Cloud cover will increase as there is more water vapour to condense. Increased cloud cover reflects more incoming short-wave solar radiation back out to space. Less incoming radiation results in temperatures falling.

At a smaller scale, feedback processes also operate within a single drainage basin. Higher levels of rainfall lead to increased evaporation, often from increased surface storage, and river discharge also increases. More water will also make its way into groundwater storage if the geology is made up of permeable rocks.

Even individual trees will show how feedback operates. In drought years, as a tree becomes more stressed due to the lack of water for its roots to take up, the tree may shed leaves in the summer to reduce water losses from transpiration. This is an example of negative feedback and is designed to ensure the tree's survival.

Feedback in the carbon cycle

At the global scale the carbon cycle is undergoing considerable changes. Rising temperatures speed up decomposition of organic material, thereby releasing more CO_2 into the atmosphere. In turn higher levels of CO_2 trap more outgoing radiation and global warming increases, which in turn speeds up decomposition and so on. This is an example of positive feedback.

Positive feedback is also seen in the Arctic. In both the Arctic tundra and the Arctic Ocean increased melting of snow and ice leads to more sunlight being absorbed as less energy is reflected back into space. The sea warms and the top layer of permafrost melts. The warmer soil then experiences higher levels of decomposition of organic material stored in it, releasing CO_2 to the atmosphere and further enhancing the greenhouse effect. There is growing concern about rising temperatures in the Arctic as so much carbon is currently stored in the permafrost, estimated at 1600 GT.

Negative feedback occurs when the higher levels of CO_2 encourage photosynthesis and growth of plants, algae and phytoplankton (primary producers). More carbon is stored in these organisms reducing CO_2 levels in the atmosphere. However, increased primary production also relies on other factors needed for photosynthesis such as sunlight, nutrients, nitrogen and water being readily available.

Exam tip

It is important to know and understand the differences between positive and negative feedback. Your responses will be more convincing if you can give examples of each.

Pathways and processes which control the cycling of water and carbon

There are various pathways and processes which control the cycling of water and carbon which vary over time.

Monitoring changes to the global water and carbon cycles

Technological advances such as computers, satellites and a great diversity of scientific instruments are generating more reliable and accurate data. At the global, continental and national scales, satellites using remote sensing have a fundamental role in monitoring changes. The growing use of Geographic Information Systems (GIS) techniques allow detailed mapping and analysis of data.

Two US-based organisations, the National Aeronautics and Space Administration (NASA) and the National Oceanic and Atmospheric Administration (NOAA) have several programmes using satellite technologies to monitor aspects such as the extent and thickness of ice-caps and glaciers, sea surface temperatures, atmospheric CO_2 and water vapour, deforestation rates and levels of primary production in the oceans. The European Space Agency (of which the UK is a member) also monitors environmental change.

The extent and thickness of Arctic sea ice is measured by satellite as well as submarine monitoring from under the ice.

Short-term changes

There are short-term changes in climate, temperature, sunlight and foliage and the water and carbon cycles.

Diurnal changes in the water and carbon cycles

Diurnal changes can be significant in the water cycle. As temperatures fall overnight, evapotranspiration reduces the flow of water to the atmosphere but

Diurnal changes take place over a 24-hour period.

then this increases as daytime temperatures rise. Convectional rainfall tends to fall during the afternoon or early evening. Direct and intense heating of a small area of the ground generates strong upward convection currents. These can lead to thunderstorms developing and are a significant part of the water cycle in tropical rainforest regions.

Diurnal carbon flows depend on whether photosynthesis is operating or not. During the day, CO_2 flows from the atmosphere to plants, while at night this flux is reversed both on land and with marine phytoplankton.

Seasonal changes in the water cycle

Seasons in any part of the world are controlled by variations in the intensity of solar radiation. In southern England, average solar input of energy is between five and six times more in June than in December. This results in high evapotranspiration rates in the summer and high flows of water from underground and the soil to the atmosphere. Stores of water reduce and river discharges are usually at their lowest.

Some parts of the world have a very clear wet and dry seasonal pattern. The monsoon regions of south and east Asia such as India and Vietnam receive high percentages (80–90%) of their annual rainfall during the three to four months of main monsoon activity. Stores fill and streams and rivers greatly increase their flow rates. Such variations can have profound impacts on agriculture and risks from water-borne diseases for example.

Where rivers start in high mountain regions, such as the Ganges in India or the Mackenzie in Canada, snowmelt in the spring and early summer can significantly increase river flows. The same is true of rivers in the Arctic tundra.

Seasonal changes in the carbon cycle

Changes in **net primary productivity (NPP)** clearly indicate seasonal change in flows and stores of carbon.

During the **growing season** NPP is greatest when carbon flows from the atmosphere to the biosphere.

In tropical rainforest areas the growing season is all year round while in the Arctic tundra it is generally between 50 and 60 days. In the northern hemisphere with its high proportion of land, ecosystem activity reduces atmospheric CO_2 by some 2 ppm during the growing season. Flowing the other way is CO_2 generated from the decomposition of organic materials.

> **Exam tip**
>
> Being able to discuss seasonal changes in the context of deciduous and evergreen forests can be helpful in understanding both water and carbon short-term cycles.

In the oceans, NPP from phytoplankton also follows seasonal variations as sea surface temperatures, more intense sunlight and increasing day length stimulate biological activity.

Knowledge check 7

Why are short-term changes to the water cycle of significance for human activities?

Net primary productivity (NPP) is the accumulation in dry weight of green plant material. NPP = gross plant production minus respiration, hence it is a net figure.

A **growing season** is the part of the year when most plant growth occurs. It is largely controlled by temperature and day length.

Long-term changes to the water and carbon cycles

Long term here means millions of years. The climate has always fluctuated over the long term. Some changes have happened relatively quickly, such as that some 65 million years ago. This was the transition between the Cretaceous and Tertiary geological periods known as the KT boundary. It is thought that a huge asteroid (10 km diameter) hit the Earth, throwing up vast quantities of dust into the atmosphere causing widespread cooling. About 15% of marine species, 75% of plant species and, most famously, the dinosaurs became extinct.

Over the last 1 million years there have been cycles of cold glacial times with warmer inter-glacials between. The last major glacial was 20,000 years ago when extensive areas of North America, northern Europe and Asia and the higher mountains of South America were covered by ice, as much as 2 km thick in some places.

Long-term changes to the water cycle

During a glacial period, much water was stored on the land as ice sheets, glaciers and in permafrost. With less water flowing round in a cycle, sea levels fell by about 100–130 metres. The loss of ecosystems meant that less water was stored in the biosphere. The climate of areas to the south and north of the ice sheets tended to become drier with deserts and grasslands replacing forests. Generally the water cycle slows during a glacial and then accelerates as ice and snow melt during an interglacial.

Long-term changes to the carbon cycle

Over the very long term, the movement of tectonic plates and changes in the rate at which carbon seeps out from the Earth's interior have altered the carbon cycle. The uplift of the Himalayas for example, beginning 50 million years ago, provided a fresh source of rock to be weathered and carbon entered the slow carbon cycle. In turn this lowered temperatures.

The carbon cycle also changes in response to climate change. During a glacial period, the carbon cycle slows. Lower sea temperatures meant that more CO_2 was absorbed by the oceans. (CO_2 dissolves more readily in cooler water.) Less atmospheric CO_2 reduced the greenhouse effect and promoted further cooling. Ocean circulation altered, bringing nutrients from the deep oceans close to the surface. In turn this stimulated phytoplankton activity which absorbed CO_2. When they died, these organisms fell to the ocean floor taking carbon into a long-term store.

Carbon storage on land decreased as tundra replaced temperate forests and grassland took over from forests in the tropics. Carbon sequestered in the soil remained locked away under ice or in the permafrost. NPP greatly reduced, as did decomposition. The overall effect was to slow the carbon cycle.

As the climate warms at the end of a glacial and the beginning of an inter-glacial, atmospheric CO_2 increases quite dramatically. NPP increases and the whole carbon cycle speeds up.

Temperate is the term given to the mid-latitude regions such as Western Europe. It indicates a climate without extremes of temperature or precipitation.

> **Exam tip**
>
> The difference between long- and short-term changes to both water and carbon cycles is important. Make sure you know what these changes are and the lengths of time involved.

The water and carbon cycles are linked and interdependent

This section will examine the extent to which the water and carbon cycles are linked and how human activities affect these linkages.

How the two cycles are linked and interdependent

How and in what ways the carbon and water cycles are linked, via the atmosphere, oceans, soils and vegetation and cryosphere, is complex.

- Atmosphere: CO_2 is exchanged between atmosphere and oceans. CO_2 contributes to the greenhouse effect. CO_2 is exchanged between soils, vegetation and atmosphere in processes such as photosynthesis, combustion, respiration and decomposition, water evaporates from the oceans and returns via various flows.
- Oceans: as atmospheric CO_2 levels rise, global temperatures rise and the ability of oceans to absorb CO_2 decreases, resulting in more CO_2 in the atmosphere. As sea surface temperatures (SSTs) rise, evaporation rates increase and more latent energy transfers to the atmosphere. Thermal expansion of sea water contributes to rising sea level. Increased global temperatures result in increased melting of ice sheets and glaciers and this water flows back into the oceans raising sea levels.
- Vegetation and soils: rates of photosynthesis, NPP, transpiration and decomposition increase when there is no shortage of water. The exchange of CO_2 with the atmosphere is therefore affected.
- Cryosphere: with increased CO_2 levels in the atmosphere, global temperatures rise causing increased melting of ice. Loss of ice (land and sea) decreases albedo and increases absorption of solar radiation increasing temperatures further. Flows of water from land to the oceans increase. Thawing of permafrost increases rates of oxidation and decomposition of organic matter, increasing release of CO_2 and CH_4 to the atmosphere where they trap yet more solar radiation.

Thermal expansion
As sea water warms its volume increases leading to sea level rise.

Human activities affect water and carbon stores

Human activities cause changes in the availability of water and carbon stores. Since all parts of the water and carbon cycles are interrelated, human activities intervening in one part of the cycles affects other parts of the cycles. Human impacts on the water and carbon cycles have grown in scale and intensity for the past two centuries, a period of significant population and economic growth.

Water stores

Irrigation of crops, increased livestock numbers, industrial and domestic use consume vast volumes of water. Water extraction from rivers and aquifers has led to severe reductions in water stores at regional and local scales. In advanced countries (ACs) sophisticated engineering systems abstract water so that river flows reduce and water table levels fall. Regional rivers such as the Colorado in southwest USA and the Po in northern Italy have been affected by low flows. In turn this can have very detrimental impacts on ecosystems.

Over-abstraction from aquifers can lead to salt water moving inland through the rocks, replacing the fresh water being removed. Such contamination has occurred along the Mediterranean coasts of Italy, Spain and Turkey, often as a result of the demands of the tourist industry (including golf courses) and agriculture. Rapidly increasing coastal communities in Bangladesh have, likewise, led to over-abstraction and salt water incursion.

The growth of urban areas increases the speed at which water enters streams and rivers. Evaporation, transpiration, throughflow and percolation are all reduced. Some places have experienced extensive deforestation, for example Amazonia, Indonesia, and parts of sub-Saharan Africa such as the semi-arid Sahel. The water cycle is thus interrupted as less water vapour returns to the atmosphere via evapotranspiration. This then reduces the probability of rainfall. In turn soils store less water and trees are unable to regenerate so both the water and carbon cycles are affected.

Carbon stores

Some carbon stores have been reduced by human activities while others have been increased. The reliance on fossil fuels over the past 200 years has led to the removal of billions of tonnes of carbon from very long-term stores (coal, oil and natural gas). At present some 8–10 billion tonnes of carbon transfer to the atmosphere from fossil fuel burning each year. The growth of economies such as China, India and Indonesia sustains this demand for fossil fuels.

Land-use changes transfer about 1 billion tonnes of carbon to the atmosphere each year, mainly due to deforestation. Increasing livestock numbers at the global scale, mainly in emerging and developing countries (EDCs) and LIDCs, also transfers large quantities of carbon to the atmosphere.

Although humans have had an impact on forests over many centuries, the intensity, speed and relative permanence of forest clearance over the past 100 years has resulted in a significant reduction in carbon stored in the biosphere. With less photosynthesis occurring, less CO_2 is absorbed by plants and this reduces the amount of carbon stored.

Perhaps even more significant is the reduction in the carbon stored in the oceans by phytoplankton. These organisms are under increasing threat due to the acidification of the oceans. Phytoplankton currently absorb over half the carbon released by fossil fuel burning which is much more than that taken up by tropical forests.

Carbon stores in wetlands drained for agriculture or building and in soils which are not sustainably managed are diminishing.

The impact of long-term climate change

Long-term climate change has an impact on the water and carbon cycles.

The water cycle

Global warming has increased evaporation, therefore more water vapour enters the atmosphere. Water vapour is a natural greenhouse gas therefore more solar radiation is trapped, raising temperatures which increases evaporation. With more water vapour in the atmosphere, precipitation is more likely. In turn this can give higher runoff of water into streams and rivers thereby increasing the flood risk.

When water vapour condenses, latent heat is given off into the atmosphere. With more energy in the atmosphere, extreme weather events such as tropical storms

> **Exam tip**
>
> For both water and carbon cycles, set your discussions clearly in the real world by quoting examples of human activities affecting the cycles and where this is occurring.

(hurricanes, cyclones and typhoons) and mid-latitude storms increase in occurrence and intensity.

Water stored as snow and ice in the cryosphere is coming out of these long-term stores as a result of global warming. The water is then transferred into the oceans and eventually into the atmosphere.

The carbon cycle

In general, higher temperatures increase rates of decomposition and so more carbon is transferred from the biosphere and soil to the atmosphere. However, higher temperatures also increase rates of photosynthesis thereby increasing carbon accumulation in plants. If rainfall patterns are disrupted by climate change, some regions may experience reduced water availability and this will cause carbon uptake by plants to diminish as vegetation patterns alter. For example, forests might be replaced by grassland in the low latitudes. In contrast, in the high latitudes global warming may allow the **boreal forests** of Canada and Siberia to expand, increasing the carbon store in these regions.

Carbon currently locked in the permafrost in the high latitudes may be released as soils thaw with global warming. The vast quantities of peat in the high latitudes, which represent a significant carbon store, could also be decomposed with higher temperatures, thereby releasing carbon into the atmosphere.

The absorption of vast quantities of CO_2 by the oceans is leading to sea water becoming increasingly acid. This causes a reduction in the phytoplankton which are responsible for storing huge amounts of carbon.

Long-term climate change may bring about an increase in carbon stored in the atmosphere, a decrease in carbon stored in the biosphere and possibly a decrease in carbon stored in the oceans. There are likely to be regional variations in carbon cycling depending on rates of photosynthesis, respiration and decomposition.

Boreal forest is the name given to the northern forests in Europe and North America. Their growing season is currently about 6 months.

Exam tip

For both water and carbon cycles you need to be able to discuss possible consequences of long-term climate change, giving evidence to support your arguments.

Water and carbon management

What are the management strategies and the global implications of water and carbon management?

Management strategies to protect the global carbon cycle

Wetland restoration

Wetlands (salt and freshwater marshes, floodplains, peatlands, mangroves) cover up to 9% of the Earth's land surface but contain 35% of the terrestrial carbon store. The common feature they share is a water table at or close to the surface, meaning that the ground is permanently saturated.

Wetlands have been drained for a wide range of human activities such as agriculture, industry and housing. This reduces habitats and biodiversity and leads to vast quantities of stored CO_2 (and methane) being transferred to the atmosphere.

Greater attention and economic value is now given to the ecosystem services benefits of wetlands, such as their role as carbon sinks. Trans-government initiatives such as the International Convention on Wetlands (Ramsar) and the European Union Habitats Directive promote restoration projects all round the world. In the UK, some 400 hectares of grade 1 farmland are being converted back to wetland in Cambridgeshire. In the Canadian prairies, over 110,000 hectares are planned for restoration to their former wetland state.

Restoring wetlands uses a variety of techniques, such as removal of flood embankments allowing water to flood back over the land, managing river levels and reducing or stopping groundwater extraction to allow water tables to rise.

Afforestation

Forests are key carbon sinks as well as having ecosystem services such as reducing flood risk and soil erosion, increasing biodiversity and regulating air quality.

The UN's Reducing Emissions from Deforestation and Forest Degradation (REDD) scheme offers incentives to countries to protect forests. The EU Forest Strategy focuses on sustainable forest management. This has involved using satellite technology to map the forest carbon store which is a prerequisite for informing decision making about how forests should be managed in different locations. In April 2016, the EU and Indonesia signed the first ever Forest Law Enforcement, Governance and Trade (FLEGT) licensing scheme. This is designed to reduce illegal logging of tropical forests and promote trade in legally produced timber. Fourteen other agreements are being negotiated with timber producing countries in southeast Asia and Africa.

Agricultural practices

Carbon is released to the atmosphere through overcultivation, overgrazing and excessive intensification of agriculture. Measures to reduce carbon emissions from agriculture target both CO_2 and methane (CH_4). They include the following.

- Zero tillage: not ploughing but drilling seed directly into the soil so conserving organic matter in the soil.
- Polyculture: growing annual crops in between trees which helps protect soils from erosion and stores carbon in the trees.
- Crop residues: leaving residues such as stems and leaves on the field after harvesting helps protect soils from erosion.
- New strains of plants, e.g. rice, which require less water in the padi fields therefore generating less CH_4.
- Managing manure, e.g. using it in anaerobic digesters to produce CH_4 which can be used to generate power.

Reducing carbon emissions

International agreements

In the 1997 Kyoto Protocol many ACs agreed to legally binding reductions in their CO_2 emissions although the USA and Australia refused to sign. Various EDCs and LIDCs, such as China and India, were exempt. The aim was to bring about a 5% cut in global GHG emissions from the 1990 levels by 2008–2012. ACs such as Japan and

Ecosystem services are the processes by which the environment produces resources used by humans, such as oxygen, water, food and materials.

Knowledge check 8

What is meant by the term 'ecosystem services'?

most of the EU were expected to cut emissions by between 6 and 8%. Negotiations continued after 1997 and although Kyoto came into effect in 2005, there was ongoing discussion and different levels of compliance amongst countries.

The Paris Climate Convention of 2015 ended with an agreement to reduce global CO_2 emissions to below 60% of 2010 levels and to restrict global warming to a 2°C increase. These aims are to be achieved by 2050. However, countries will set their own voluntary targets and there is no detailed timetable. It has been agreed that ACs will transfer substantial funds and technologies to assist EDCs and LIDCS to achieve their targets.

EDCs and LIDCs argue that their current economic and social situations necessitate considerable use of fossil fuels to power an increase in living standards their people deserve. It is argued that present-day ACs have achieved their high standards of living through fossil fuel combustion over the past 200 years so it is they who are primarily responsible for the increases in greenhouse gases (GHGs) such as CO_2.

Cap and trade

Kyoto brought in the idea that a country which cut its emissions below the level it had been 'capped' at, would be able to trade the unused volume of emissions. Another country could buy these theoretical emissions and offset them against emissions above the agreed level. This would allow it to achieve its target or cap.

The EU introduced a similar cap and trade system in 2005 with individual businesses, especially energy intensive ones, such as metal, cement or refining industries, able to receive credits if they achieved lower-than-set emissions, which could then be sold. The idea of this carbon market is that polluters have to pay while clean companies are rewarded.

Carbon offsets are credits for schemes which promote carbon sequestration such as afforestation or reduce emissions, such as using renewable energy.

Another Kyoto initiative is the Clean Development mechanism. ACs with a GHG reduction programme can invest in emission-reducing projects in EDCs or LIDCs. These investments in projects such as wetland restoration and forest conservation are taken as alternatives to expensive emission-reduction projects in the ACs.

Management strategies to protect the global water cycle

Forestry

Forests play a vital role in the water cycle as well as in the carbon cycle. Strategies involving forestry, such as the REDD programme, therefore protect both cycles.

Tropical forests are given much attention. The Amazon Regional Protected Area (ARPA) strategy receives support including funding from a variety of sources such as the UN, the World Bank, the EU, the German Development Bank and NGOs, for example the World Wildlife Fund (WWF). Almost 10% of the Amazon basin is protected from activities such as logging, ranching and mineral exploitation.

Exam tip

Knowing details about various international agreements, such as dates and emission levels aimed for, makes your response more authoritative.

Exam tip

Be up to date with what is happening as regards international agreements aimed at reducing carbon emissions. Try to identify examples of practical projects which are actually happening, not just wishful thinking. Be able to offer authoritative assessment of their likely impacts on emissions.

Water allocations

Fresh water is a renewable resource but such is the increase in the rate at which it is withdrawn for human activities, that issues about water availability and quality have become acute, especially in semi-arid and arid regions.

Agriculture is the biggest consumer of water accounting for just under 75% of global water withdrawals (water taken from rivers, lakes and aquifers). Irrigation of crops can lead to much water being lost to evaporation, such as when sprayed onto a crop, or percolation when too much water is used in a field. Strategies to make water use more efficient include:

- drip irrigation which leaks water from pipes running in or on the soil
- mulching of crops with organic matter which reduces evaporation and increases water storage in the soil
- crop breeding to develop varieties of plants with lower water demand.

Agriculture that uses steep slopes, terracing, contour ploughing and strips of vegetation reduces surface runoff and therefore promotes infiltration and water storage in the soil. Increased water harvesting and storage by farmers, such as collecting water off barn roofs and storage in giant tanks or specially dug ponds and reservoirs, helps manage the water.

As with carbon, water is increasingly being traded in markets. In Queensland, Australia, active water markets exist. The aim is to encourage users, such as agricultural, industrial or domestic consumers, to appreciate water as being of high value and therefore to use it sustainably. In the UK, building regulations now state that water-saving devices in items such as toilets must be installed in new-build housing. Increasing numbers of users install water meters in order to help them manage their water consumption.

Drainage basin planning

The drainage basin scale is the most effective unit in which to manage the water cycle. The inputs, stores and outputs can often be managed effectively as an open system across all the users operating in the drainage basin.

In England and Wales, several river basin districts have been identified. These vary from those focused on one major river, such as the Severn, to the South West which consists of several river basins such as the Exe, Stour and Tamar. Each basin has a management plan with the purpose of providing a framework to protect and enhance the benefits available from the water cycle operating in the basin. Because water and land resources are so closely linked, the management plan covers not only matters such as the volume of water abstracted and water quality but also land-use decisions such as housing developments.

Where a river flows through more than one country, management can prove to be extremely complex. Transboundary issues tend to slow down agreements as to how the water cycle should be used or make them virtually impossible. The Mekong in southeast Asia travels through six independent states. In China, the river is used for hydro-electric power generation which has knock-on effects downstream in countries such as Lao PDR (Laos) where fishing is important. Geo-politics can be a significant element in managing river basins in some regions.

Summary

- Water and carbon are vital for life on Earth.
- Water and carbon move between the land, oceans and atmosphere through open and closed systems.
- The water and carbon cycles operate as systems with inputs, stores and outputs.
- There are distinctive processes and pathways of water and carbon in their cycles.
- Case studies are required of two contrasting locations, a tropical rainforest and the Arctic tundra, to illustrate their water and carbon cycles, physical and human factors causing change to the cycles and strategies to manage the water and carbon cycles.
- Human factors can disturb and enhance natural processes and stores in the water and carbon cycles.
- There are short-term and long-term changes to the water and carbon cycles caused by both natural and human factors.
- There are various techniques to research and monitor changes in the water and carbon cycles.
- The water and carbon cycles are linked and interdependent.
- Global management strategies aim to protect the water and carbon cycles.

◼ Global connections

Trade in the contemporary world

This section examines the contemporary patterns of international trade.

Flows of merchandise, services and capital

The global pattern of international trade is uneven. It is dominated by **advanced countries** (ACs) such as the USA and the faster growing **emerging and developing countries** (EDCs) such as China. **Low income developing countries** (LIDCs) such as Sierra Leone have limited access to international markets and as a result have relatively weak **terms of trade**.

Spatial patterns of international trade are complex and are better understood by considering the three main components of the **global trade system**. These include **merchandise**, **services** and **capital,** each of which incorporates a wide range of products. These products are traded at different scales, between countries, regions and continents. The flows of these items within the global trade system vary in direction, composition, volume and value.

Merchandise, services and capital as components of international trade

Merchandise

Merchandise in the global trade system can be defined as all **commodities** or products (not services) that move through a country either as imports or exports. The World Trade Organisation (WTO) identifies three categories of merchandise: manufactured goods, fuels and mining products, and agricultural products.

> **Exam tip**
>
> Appropriate use of key terms such as 'terms of trade', 'balance of payments' or 'economic multiplier', is important in all answers. In the shorter data-response sections this allows your responses to be precise and concise.

A feature of the spatial pattern of merchandise trade is the large inequalities in the trade of these primary and secondary products. For example, the value of exports of merchandise for Europe and Asia is much greater than that of Africa, especially in manufactured goods.

The magnitude of global inequalities in merchandise trade is illustrated by the contrasting statistics for the USA and Sierra Leone (Table 5).

Table 5 Contrasts in merchandise trade for the USA and Sierra Leone, 2014

		USA	Sierra Leone
World ranking for merchandise trade	Exports	2	139
	Imports	1	163
Value of merchandise traded (US$ million)	Exports	1,620,532	1,886
	Imports	2,412,547	1,489
Share in total world trade of merchandise (%)	Exports	8.53	0.01
	Imports	12.64	0.01

These inequalities are explained by a combination of economic, social, environmental and political factors (see Figure 5 on page 36). For example, the USA's trade strength is the product of its ability to:

- invest in transport infrastructure, technology in communications, domestic industry, and outward FDI
- exploit its wide range of natural resources including minerals and ores, climate and coastline with natural harbours
- negotiate trade agreements, gaining relatively easy access to global markets
- employ a highly skilled and educated workforce.

Sierra Leone has limited access to global markets because of factors such as its:

- overdependence on primary product exports which are vulnerable to economic shock
- inadequate infrastructure and poor communications networks
- slow recovery from the consequences of civil war
- high levels of unemployment and an inadequately educated workforce
- restrictive human rights issues such as gender inequality and use of child labour
- high levels of poverty, low life expectancy and the effects of many prevalent diseases.

Global trade in oil illustrates the idea of flows of merchandise within the global trade system. The spatial patterns of flow are intricate simply because there are many areas of supply and demand. The pattern is influenced by factors which include: the location of the oil fields, the ability of a country to exploit its oil reserves, the level of investment in technology for exploration and extraction, availability of a skilled workforce, the ability to export oil to a variety of global markets via appropriate means of transport, the location of areas of demand for the oil, government policy, and the regulation of organising bodies such as OPEC (Organisation of Petroleum Exporting Countries). The flows of trade in oil also vary over time and depend on factors such as economic and political shocks which can affect supply and the world price of oil.

Services

Commercial services include a wide range of products in many fields including transport and travel, communications, construction, insurance, finance, ICT, and government services. Globally the value of exports of services is much smaller than that of merchandise but this is a rapidly growing element of the global trade system.

A feature of the spatial patterns of trade in commercial services, like merchandise trade, is the great size of inequalities at regional and national scales. For example, Europe is the largest net exporter of commercial services and Africa is the largest net importer. The largest exporters of commercial services are found in the EU, North America, Japan and emerging economies such as China and India. These contrast significantly with the very low figures for the countries of sub-Saharan Africa.

It must be remembered that these global regions are very large areas containing many variations between countries. But overall it is the ACs which have the strongest **balance of payments** as a result of commercial services transactions as shown by the USA and Sierra Leone contrasts (Table 6).

These global contrasts in ability to supply commercial services depend on factors such as:
- government and private investment in service industries, including the outsourcing of services
- skill and education levels of the workforce
- levels of investment in communications, including ICT, and transport
- the strength and reliability of financial and legal services.

Table 6 Contrasts in trade in commercial services for the USA and Sierra Leone, 2014

		USA	Sierra Leone
World ranking for trade in commercial services	Exports	1	166
	Imports	1	151
Value of trade in commercial services (US$ million)	Exports	687,605	191
	Imports	451,683	640
Share in total world trade of commercial services (%)	Exports	13.92	<0.01
	Imports	9.44	0.01

Flows of capital within the global trade system involve many diverse elements. These include:
- the purchase of both real and financial assets — real assets include tangible items such as merchandise and property, and financial assets or intangibles include currency, stocks and bonds — and of course any transaction involving these assets generates a counter flow of finance
- international trade in foreign exchange reserves
- **foreign direct investment** (FDI).

Spatial patterns of these transactions reflect the increasing interconnectivity of global trade. Most flows are between a small number of ACs, but increasingly EDCs and LIDCs are becoming integrated into the global trade system which involves flows of capital.

Knowledge check 9

Explain the difference between 'terms of trade' and 'balance of payments'.

Knowledge check 10

What is meant by FDI?

Much of the flow of capital in the global trade system is intra-firm — that is, it takes place within large companies as their international trade flows along the **global supply chains** which they have established. In this respect it is the large **multinational corporations** (MNCs) which dominate international trade.

Foreign direct investment

A key element of global capital flows is the direct investment by companies in foreign countries. Much of this FDI is conducted by the many MNCs. And most of this investment emanates from ACs where the MNCs have their main company headquarters. They invest in subsidiary or component factories in EDCs and in LIDCs or in other ACs. This is a long-established form of investment which has continued to grow as economic globalisation proceeds, creating links between the economies of all types of country in the development spectrum.

The Indian government has actively encouraged **inward FDI** through its own sponsorship, for example by establishing treaties with Mauritius and Singapore. Large companies locate in these countries and avoid extra taxation when they further invest in India. The Reserve Bank of India assists in the necessary financial transactions. The investment has been very strong in manufacturing, construction, energy generation and communication services. The main reasons for this encouragement are to develop infrastructure (ports, airports and highways), defence capability, telecommunications and pharmaceuticals industries in India. In addition through **outsourcing** there has been significant inward investment into India by insurance and finance companies, and business and computer services. Much of this investment is from large MNCs based in ACs.

Increasingly the global share of FDI is being derived from EDCs themselves. China is one of several emerging economies with **outward FDI** interests in sub-Saharan Africa. The others include Brazil, India, South Korea, Russia and Turkey. Of these China is the largest EDC investor in Africa, including both government sponsored and private investment flows. Its investments have risen rapidly from US$15 billion in 2005 to over US$100 billion in 2013. Chinese global investments have been spatially widespread and in many diverse sectors.

In sub-Saharan Africa for example there has been considerable Chinese outward FDI in:

- minerals in countries such as Angola, Chad, Niger, Nigeria, Sudan and Zambia
- manufacturing in Ethiopia (glass, fur, footwear, automobiles), Mali (sugar refining) and Uganda (textiles and steel pipes)
- regional rail networks in east Africa.

Advantages for China are that **global value chains** are established and costs are reduced as its own domestic labour costs begin to rise. In return, the sub-Saharan African countries develop their trade by gaining access to Chinese markets, many benefiting from zero tariffs.

Current spatial patterns in international trade

This section will examine the current spatial patterns in the direction and components of international trade.

Exam tip

Ability to write clear definitions of geographical terms such as 'the development gap', 'economic interdependence' or 'global supply chain' can be helpful in essays. This demonstrates your understanding of key concepts which are important in the discussion.

Exam tip

You should be prepared to identify features of spatial patterns shown on different types of statistical maps such as choropleths or flow line maps. This could involve global or regional patterns of trade in specific types of merchandise, services or capital.

Knowledge check 11

What is meant by 'spatial pattern'?

Inter-regional trade between Europe and North America

There are long-established trade relationships between Europe and North America. An example of a current agreement being negotiated between the EU and the USA is the Transatlantic Trade and Investment Partnership (TTIP). This has removed trade barriers by lowering tariffs, which has helped to promote trade and encourage economic growth. As a member of the EU, the UK benefits from this partnership and it also has its own bilateral trade agreements with each of the three NAFTA members (USA, Canada and Mexico).

There is a strong degree of economic interdependence between the UK and USA. The UK is the main single country export market for the USA and vice versa. Trade in merchandise between the UK and USA is just one aspect of their **economic interdependence**. The main exports from USA to UK and from UK to USA are from similar sectors of manufacturing. Machines, engines, pumps, vehicles, aircraft, spacecraft and pharmaceuticals are all examples of merchandise traded in both directions across the Atlantic between the two countries on a reciprocal basis. These include finished products, such as different types of cars, or component parts of goods such as vehicles or spacecraft. These are goods in which each of the UK and USA specialise, or are most efficient at manufacturing — that is, goods which each country has greatest relative advantage in producing. This particular example of inter-regional trade in merchandise illustrates the principle of comparative advantage.

The inter-regional trade partnership between the UK and the USA is further exemplified by the high value in flows of FDI between the two countries. The figures vary annually, but the UNCTAD database shows millions of dollars of outflows invested by UK companies in the USA and by USA companies in the UK each year. This contributes significantly to economic growth and employment in both countries.

Intra-regional trade within the EU

Most international trade is intra-regional. The EU is a large **trading bloc** with 28 member states. There is a huge volume of trade between these countries, not only in merchandise but also in commercial services and capital. According to Eurostat statistics, in 2013 the total value of intra-EU trade in merchandise was €2,935 billion and in services it was €842 billion.

Patterns of intra-regional trade in the EU are complex. This can be explained by economic and geographical factors such as: the high degree of interconnectivity of transport networks within the EU, the relative ease of financial transactions especially within the Eurozone, the single market economy, the trade defence policy of the EU which helps to protect domestic industries, the lack of internal tariffs between the 28 states, and the increasing number of EU based MNCs which operate within the trading bloc.

In addition there is a wide range of raw materials, minerals, agricultural products, manufactured goods and commercial services which can be produced and supplied within this geographically diverse region, and there is significant demand from its 520 million population.

The UK food and drink sector demonstrates how intra-EU trade in merchandise is affected by a combination of these factors. In 2014 the UK exported 75% of its food and drink products to other EU countries. These included milk to Ireland, salmon to

Inter-regional trade is the flow of international trade among major world regions.

Comparative advantage is the principle that countries or regions benefit from specialising in an economic activity in which they are relatively more efficient or skilled.

Intra-regional trade is the flow of international trade within one or other of the major world regions.

A **single market** is an economic union of countries trading with each other without any internal borders or tariffs.

France, confectionery to the Netherlands and wheat to Spain. Trade in these goods can be attributed to physical conditions and human expertise needed to produce quality food and drink, government and private investment in marketing, free trade within the EU and the technology to transport these perishable goods over long distances.

Romania joined the EU in 2007. Since that time, for these same economic and geographical reasons, the total value of its exports has increased each year and its trade relations have strengthened. In 2015, its biggest trading partners were Germany, Italy and France — all EU members.

Factors that influence patterns of international trade

The global pattern of international trade is one of great inequality. It is dominated by ACs which are in a strong position to drive their economies to their own advantage, whereas many peripheral LIDC economies have limited access to global markets and are able to participate in global trade in only a restricted way. Nevertheless there is change within this pattern as EDCs improve their trade strength and as globalisation proceeds. Factors that influence this pattern and underpin these inequalities with respect to merchandise trade are categorised as physical, economic, social and political (see Figure 5).

ECONOMIC	POLITICAL
Physical infrastructure including transport Technology including communications Transport costs Cost of production Foreign direct investment Speed of border formalities, e.g. documentation / use of ICT	Supra-national organisations Regional trading blocs National government policy Trade agreements / market integration Tariff / non-tariff barriers Free trade / Free trade areas Governance / transparency of customs authorities
SOCIAL	**ENVIRONMENTAL**
Demographic factors which affect labour force and import demand including age structure and migration Stage in demographic transition Female empowerment / women in the labour force Levels of education	Distribution of natural resources including oil and mineral ores Climate / soils / water scarcity (affecting food and agricultural products in international trade) Deep-water ports Natural hazards

Figure 5 Factors influencing contemporary patterns of international trade

International trade and socio-economic development

The relationship between current patterns of international trade and socio-economic development

There is a strong positive statistical relationship between the value of exports of countries and their **Human Development Index (HDI)**. ACs, such as the USA, have high total export values and high HDI, whereas LIDCs, such as Haiti, have

Exam tip

Ability to describe the relationship between two sets of data with reference to a scatter graph, a table or other statistical diagrams is useful. You should be able to explain the relationship and any anomalies, such as that between percentage share of global merchandise exports and HDI.

low total export values and low HDI. Anomalies include New Zealand which has a relatively low total export value and high HDI and China which has a relatively lower HDI than might be expected for a country with such a high total value of exports.

Value of exports and HDI are two examples of indices used to assess the nature of the relationship. Others include specific indices of international trade, such as value of outward flow of FDI or value of commercial services exports, and indices of socio-economic development such as infant mortality rates or maternal mortality rates.

The relationship is causal. The impact of international trade on socio-economic development is widely recognised, not least by the World Trade Organisation (WTO) and national governments. Ways in which trade influences development are outlined below and in the guidance for case studies later in the chapter.

International trade promotes stability, growth and development

The link between international trade and development is demonstrated by the effect of international trade in promoting stability, growth and development within and between countries.

Stability

Trade can contribute to stability in the following ways.
- Trade can contribute to international peace and stability, especially if countries trade under the same rules, for example under the 'most-favoured-nation' principle established by the WTO.
- Trade encourages states to cooperate. Multilateral and bilateral trade agreements can contribute to economic and political stability.
- Some bilateral agreements extend beyond trade, and may lead to cooperation and assistance in dealing with political issues such as strengthening democratic processes and human rights, which create a more stable environment for foreign investors.

Economic growth

Trade can contribute to economic growth in the following ways.
- Trade in merchandise and commercial services stimulates production, contributes to GDP growth and to further investment, including FDI.
- Employment opportunities are created, incomes are raised and in some LIDCs and EDCs poverty levels can be significantly reduced.
- The **economic multiplier** can be enhanced by international trade. This can bring benefits at different scales from local communities to urban and regional economies.

Development

Trade can contribute to development in the following ways.
- Removal of tariffs and other obstacles to LIDC trade helps to generate foreign exchange which can be invested to reduce internal inequalities in poverty, health, education, infrastructure and transport.
- The Corporate Social Responsibility of MNCs can be of economic and social benefit to employees and communities in areas of production.

Knowledge check 12

What is meant by HDI?

Exam tip

Responses to questions on the impact of international trade should consider different perspectives. These impacts may be viewed as: economic, social, political or environmental; positive/advantages or negative/disadvantages; short- or long-term; and local scale through to global scale.

Knowledge check 13

What is meant by 'the economic multiplier effect'?

Corporate Social Responsibility is the commitment and initiative of a corporation to assess and take responsibility for its social and environmental impact. This includes its ethical behaviour towards the quality of life of its workforce, their families, and local communities, and its contribution to economic development and the natural environment.

- Membership of regional trading blocs and political unions which have a common purpose can help socio-economic development within member states.

Flows of people, money, ideas and technology are linked to these positive effects of international trade. Examples include the following.

- Migration of highly skilled workers such as scientists and engineers can be innovative in circulating ideas and information on technology development between ACs and EDCs. Migrants returning to LIDCs from ACs are able to contribute newly acquired skills and knowledge.
- Foreign exchange generated by international trade is the monetary flow which can stimulate further domestic and foreign investment.
- Trade agreements can lead to acceptance of human rights, for example in negotiations to combat child labour in many **free trade agreements (FTAs).**
- Global spread of ICT has helped speed up financial transactions and movement of merchandise through customs, reduce corruption, improve the logistics of tracking products in supply chains, and improve cyber-security.

International trade causes inequalities, conflicts and injustices

International trade causes inequalities, conflicts and injustices for people and places. Rich and powerful countries dominate international trade. This can cause inequalities, conflict and injustices which have an impact on poor countries in particular.

Inequalities

International trade causes inequalities in the following ways.

- Many LIDCs have limited access to global markets. This widens the **development gap** between developed and developing countries.
- Skilled workers, especially men, tend to benefit most from employment opportunities created by trade, whereas many unskilled workers and women held back by limited education opportunities remain unemployed and unable to contribute to the workforce.
- In many LIDCs internal inequalities are exacerbated by trade activity, often spatially concentrated in ports where most commercial activity is located.

Conflicts

Conflicts can also occur as a result of international trade.

- Trade disputes can arise over tariffs, prices of commodities and changes in trade agreements.
- Border and customs authorities can be subject to corruption and breaches of security.
- Port development, mining and deforestation linked to trade create environmental conflicts.

Injustices

International trade can also cause injustices.

- Displacement of communities can result from land grabbing by investments in industry and agri-business.
- Attempts to secure cheap labour can result in the use of child labour and other forms of modern slavery.

Exam tip

Essay questions requiring evaluation of trade related issues should have a discursive response that includes a range of arguments, a balanced assessment of the issues, and a conclusion consistent with that balance.

Knowledge check 14

What do you understand by the development gap?

- Unequal power relations, unfair trade rules such as tariffs and other trade barriers and opening up to free trade can adversely affect businesses such as small-scale farmers or fishermen in LIDCs.

Unequal flows of people, money, ideas and technology are linked to inequalities, conflicts and injustices, such as the following.

- Access to technology is unequal across the world. Examples include the huge differences in investment in new technology for handling large-scale shipping, or in internet and mobile phone subscriptions.
- Foreign investment in mining operations can have negative effects on indigenous populations such as displacement from their traditional homeland areas and damage to natural environments.
- Investments in localised areas such as ports in LIDCs can lead to internal migrant flows as people seek employment opportunities. This contributes to the widening of socio-economic inequalities within a country.

Why has trade become increasingly complex?

Factors influencing access to markets

Access to markets is influenced by a multitude of interrelated factors. International trade has increased connectivity due to changes in the twenty-first century.

Technology, transport and communications

Supply chains

Global supply chains form a network of manufacturers, suppliers, distributors and customers. Materials, products, information and finance flow within this network. The network is spreading geographically as more countries and corporations become interconnected within the global trade system.

Supply chains are of fundamental importance in international trade. Integration into supply chains has been made easier by aspects of globalisation such as the increasingly widespread use of ICT and the increasing investment in transport and trade-related infrastructure.

The term **global value chain** is used to indicate where value is added to merchandise in the supply chain at each stage of production.

Companies involved in global supply chains are subject to the possibility of different types of risk, including the quality of the product itself, political instability, economic instability or shock, the effects of terrorism and piracy, cyber-crime, ethical issues such as use of child labour, customs issues, and the effects of natural hazards.

Communications and technology

Digital connectivity is important in the development of supply chains: it helps to connect producers and customers quickly and easily, it is important in the delivery of services especially finance, and it plays a significant part in the work of customs agencies to ensure efficient administration and governance of corruption.

Exam tip

Be prepared to identify patterns and trends over time from statistical tables and graphs such as line or bar charts. This could include data on changes in value of merchandise exports, flows of FDI, or changes in indices of development such as HDI for particular countries.

Knowledge check 15

What are global supply chains?

There is close correlation between global patterns of access to broadband, use of ICT, and patterns of international trade. For example, the contrast between sub-Saharan Africa and Europe in the development of communications and technology is significant in explaining contrasts in the volume and value of international trade between these two regions.

Transport and technology

Investments in transport infrastructure and transport technology have been important in improving connectivity within supply chains. This includes:

- development of deep-water ports, essential for the import and export of most goods by container, with berthing facilities capable of handling large ocean-going ships
- investment in computer systems which deal with the logistics of container distribution
- expansion and modernisation of domestic road, rail and air transport infrastructure, essential for moving goods within a country.

There are significant differences in these types of investment between ACs and LIDCs. For example, investments in expansion of deep-water container ports such as Felixstowe and London Gateway, in transport networks in the hinterland and computerised logistics are a feature of only a limited number of ports throughout the world, mostly in ACs and some EDCs (Figure 6).

Figure 6 High levels of investment in transport and technology in an advanced country at the Port of Felixstowe, UK. This illustrates one of many sets of factors that influence access to global markets and patterns of international trade

Increasing influence of MNCs in EDCs

MNCs dominate international trade. These large companies are a powerful economic force and main drivers of the global trade system. Most have headquarters in ACs, especially in Europe and the USA. They are responsible for developing global supply chains through investments in factories and businesses in EDCs, such as India and China. This trend of economic globalisation has continued in the twenty-first century with increasing investments of AC-based MNCs in sub-Saharan Africa, southeast Asia and Latin America.

MNCs can bring social, economic and environmental advantages and disadvantages to countries in which they invest often at a local scale. Benefits include employment opportunities, higher incomes and stimulation of the economic multiplier effect. Organisations such as OECD and the UN attempt to reinforce Corporate Social Responsibility amongst MNCs in order to reinforce the benefits and to minimise the problems such as workforce exploitation and environmental pollution.

A significant new trend in the twenty-first century which has added to the complexity of international trade has been the level of MNC investment emerging from the EDCs themselves. For example, China's outward FDI has increased five-fold in the last decade (Figure 7). These investments, especially in sub-Saharan Africa, are now not just exploitative for primary products but are increasingly in secondary and tertiary sectors.

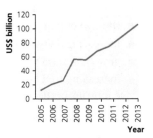

Figure 7 China — outward foreign direct investment 2005–2013

Outsourcing

Outsourcing is a cost saving strategy where a company, which has comparative advantage, provides goods and services for another company even though they could be produced in-house. This is a long-established form of investment but it continues to grow. Most large corporations invest in outsourcing of services. Many outsourcing destinations for ICT services are in India where there are government incentives for investment, lower labour costs, and a highly skilled professional workforce in ICT in locations such as Bangalore.

Role of regional trading blocs

Regional trading blocs are groups of countries which are geographically close to each other and form free trade areas, customs unions or economic unions, for example the European Union (EU), the North American Free Trade Agreement (NAFTA) and the Association of Southeast Asian Nations (ASEAN).

The trade policies of regional trading blocs provide economic benefits for member states and as a result they have become increasingly powerful organisations in the global trade system. The high volume of **intra-regional trade** generated within them between the member states makes them accountable for most of global international trade. Their negotiating strength has strong influence on the direction, composition and volume of their **inter-regional trade** with other countries and trading blocs.

The EU is the largest bloc, with 28 member states. Trade between these countries alone accounts for 16% of total world trade. In 2013 the EU was the largest trade partner of 59 other countries outside the EU.

The EU has achieved this strong global trading position through its trade defence policy, its trade and development policy, its negotiation of free trade agreements and its development and trade partnerships.

Growth of 'South–South' trade

South–South trade is the trade between developing countries which includes both LIDCs and EDCs. The rapid increase in international trade between countries of the economic 'south' has been a driver of growth for many developing countries. This trend has been enhanced by:

- demand for raw materials and energy in China and India
- the size of potential markets in Asia and Latin America
- increasing demand from the growing middle class in Brazil, India and China
- growth in intra-regional trade within trading blocs in the 'south'
- increasing FDI between EDCs such as China and India and other developing countries in the 'south'.

These are significant factors in explaining the increasing complexity of international trade.

Despite the many barriers to international trade and the difficulties of integration into supply chains, even the relatively poor, marginalised countries of sub-Saharan Africa are beginning to benefit from both intra-regional trade such as within ECOWAS, and also inter-regional trade with other LIDCs and EDCs such as China.

Growth of services in the global economy

In the twenty-first century, global trade in commercial services has expanded rapidly. Europe and North America were the largest exporters of services in 2013: it is the wealthy advanced economies which are in a strong position to drive this type of international trade. But their global share has been dropping in recent years as their former markets in Asia have become service exporters themselves. This is largely the result of the emergence of ICT services and growth in travel services to new tourist destinations.

Economic growth in the BRICS (Brazil, Russia, India, China and South Africa) countries has led to significant growth in their ability to provide commercial services. These include services in financial, ICT, communications and transport sectors. This growth in commercial services has added to the interconnectivity of international trade.

According to the International Monetary Fund (IMF), income generated from growth in trade in commercial services is essential for sustained economic development, especially in LIDCs.

Increasing labour mobility and the new international division of labour

The term **new international division of labour (NIDL)** describes the reorganisation of production at a global scale which has occurred over the last 30–40 years as a result of de-industrialisation in advanced countries and the global spread of MNCs. This has produced a broad pattern of higher-paid managerial jobs and research and development in ACs and lower-paid labouring jobs, mainly in construction and manufacturing, in LIDCs.

This change is part of the economic globalisation process as MNCs have developed supply chains in the ever widening global network of international trade. For example, new inroads of investment are now being made into many of the sub-Saharan LIDCs. NIDL has therefore contributed to the increasing complexity of international trade.

> **Exam tip**
>
> When giving explanations of spatial patterns or trends over time for trade data, refer to the influence of specific types of physical, economic, social and political factors. Also demonstrate your understanding of how these factors are interrelated or apply in combination.

> De-industrialisation is the absolute or relative decline in the importance of manufacturing in the economy of a country or region.

Over the same period, this shift in industrial structure has influenced the movement of labour at all scales, for example from rural to urban, to intra-regional and inter-continental. In the twenty-first century this has been accompanied by significant increase in labour mobility. There has been an increasing range of countries of origin and destination for economic migrants. Factors which have contributed to this growth include the following.

- Ease of labour migration within large trading blocs such as ASEAN, MERCOSUR and the EU. The large scale of labour migration within these regions is driven by inequalities in income and variations in employment opportunities for low-skilled labour between member states.
- Ease and reliability of making transfers of migrant remittances via relatively secure financial programmes and increasing availability of ICT and mobile internet subscriptions.
- Immigration policies of individual governments attract highly skilled and/or low-skilled workers.

Knowledge check 16

Give three examples of physical factors that can influence spatial patterns of merchandise trade.

Interdependence between countries and their trading partners

Case study of one EDC

A feature of the intensification of globalisation in the twenty-first century has been the increasing **interdependence** of trading nations. More countries, especially rapidly growing EDC economies, have become interconnected and integrated into the global trade system.

To illustrate this topic, and to show full understanding of the term interdependence, the specification requires a case study of an EDC such as India, Brazil or China.

For your chosen case study you should be able to illustrate the following.
- The direction and components of its current international trade patterns. These could include trading partner links with examples of imports/exports of merchandise, services and capital.
- Changes in its international trade patterns over time as the country has developed. These may be explained by factors such as government trade

policy, any change from import substitution to a more liberal policy (India), development of a wider range of trading partners and foreign investors.
- Economic, political, social and environmental interdependence with trading partners. This could include bilateral trade relationships, plus the strengthening of political relationships, especially with neighbouring EDCs or other partners within a trading bloc that experience similar problems and share similar aims for development. This might also include the increasing cooperation with neighbouring countries over environmental issues or social relationships in terms of diaspora populations.
- Impacts of trade on the EDC, including economic development, political stability and social equality. This could include data on specific indices such as changes in HDI or inequality measured by the Gini coefficient and should include an understanding of the factors which underpin these impacts.

Knowledge check 17

In terms of international trade, what do you understand by the economic interdependence of countries?

Exam tip

Interdependence between countries is developing through international trade and is an important aspect of globalisation. You should understand that interdependence has economic, social, political and environmental dimensions.

The issues associated with unequal flows of international trade

International trade creates opportunities and challenges

The economic and political strength of ACs in the global trade system creates many benefits for countries such as the USA, UK or Japan. But it can also lead to problems such as border control, the effects of trade deficit and environmental impacts. And while international trade contributes to socio-economic development in LIDCs, their limited influence in the global trade system leaves them with challenges such as integration into global supply chains, attracting investment and dealing with internal inequalities. The opportunities and challenges which arise from these global inequalities in international trade which reflect the unequal power relations between countries can be exemplified by case studies of an AC and an LIDC.

Exam tip

You will need to know a range of opportunities and challenges for people and places resulting from international trade. Make sure you can write about these in economic, social and environmental terms for the places you have investigated.

Case study of one AC

This part of the specification requires a case study to illustrate how **core economies** have a strong influence and drive trade in the global system to their own advantage. Your case study should be an AC such as the USA and you should be able to identify the combination of factors which explain the following.

- The components of its international trade, its patterns of trade, trade partners and trade negotiations and agreements. These are likely to include the main sectors of international trade – merchandise, services and capital, the benefits of bilateral trade partnerships negotiated with other countries and trade agreements with trading blocs, and intra-regional trade through membership of a trading bloc.
- Its advantages for international trade, which could include a range of economic, social, political and environmental factors, such as: investment in domestic transport infrastructure, industrial productivity, outward FDI, technology

in transport and communications, ability to exploit natural resources, political strength in negotiating trade agreements, levels of skill and education in the workforce.
- Opportunities that international trade creates for the chosen AC. These might include: employment opportunities in a wide range of industrial sectors, stimulation of the economic multiplier effect at various scales, development of positive political and cultural relationships with its trade partners including stewardship of the environment, creation of sustained economic growth, and ability to integrate other countries, such as LIDCs, into global supply chains.
- Challenges which arise as a result of its international trade, including: local pollution issues and land-use conflicts, for example resulting from port development, trade disputes over 'price dumping', managing any trade deficit, customs security and border control, and how it builds resilience to withstand economic shock.

Exam tip

You should be able to explain the unequal power relations within the global trade system, and the issues arising from these inequalities for one AC and one LIDC.

Knowledge check 18

Identify three socio-economic opportunities that international trade might create for an LIDC.

Case study of one LIDC

This part of the specification requires a case study to show how **peripheral economies** have limited influence and can only respond to change in the global trade system. Your case study should be an LIDC, such as Sierra Leone, and you should be able to identify the combination of factors which explain its relatively weak position in international trade, including the following.

- The components of its international trade, its patterns of trade, trade partners and trade negotiations and agreements. These are likely to include the main sectors of international trade – merchandise, services and capital, the benefits of bilateral trade partnerships negotiated with ACs perhaps in other regions, and any intra-regional trade through membership of a trading bloc.
- Its limited access to global markets resulting from a range of economic, social, political and environmental factors. These could include: the country's limited ability to exploit, transport, market and export its primary products, its ability

to cope with economic shock such as changes in global demand and prices for primary products, its vulnerability to natural hazards, and the effects of political shock on its economy such as conflict.

- Opportunities brought by international trade, which could include: economic development, diversification of industry, investment in infrastructure, socio-economic development through investment in health and education, improvements in financial transactions, customs security, human rights, crime and conflict through relationships, and interdependence with countries such as the UK and USA.
- Challenges which remain such as: achieving political stability and democracy, removing barriers which prevent integration into global value chains, such as illegal practices or limited investment in transport infrastructure, managing environmental problems arising from mining operations, and reducing socio-economic inequalities.

Summary

- The global trade system involves international flows of merchandise, services and capital.
- Contemporary spatial patterns of international trade are influenced by a combination of economic, political, social and physical factors.
- Spatial patterns of international trade are closely related to global patterns of socio-economic development. Global trade is dominated by advanced countries (ACs) and the faster growing emerging and developing countries (EDCs). Many low income developing countries (LIDCs) have weaker terms of trade and are peripheral within the global trade system.
- International trade can promote stability, economic growth and socio-economic development. It can also have the effect of causing inequalities, conflict and injustice.
- The global trade system incorporates flows of people, money, ideas and technology. These all have an impact on international trade and the development process.
- International trade has become increasingly complex. Access to global markets has been influenced by a range of interrelated factors. These

have led to change in the twenty-first century and include: the effects of technology, transport and communications on global supply chains, the increasing influence of MNCs, including those emerging from EDCs, the role of regional trading blocs in international trade, the growth of South–South trade, the growth of trade in commercial services, and the increase in labour mobility and effects of new international division of labour.

- There has been increasing interdependence between countries and their trading partners within the global trade system. Economic interdependence is sometimes extended to social, political and environmental interdependence as relationships between countries develop as a result of continued globalisation in the twenty-first century. This can be exemplified by a case study of an EDC such as India.
- International trade creates both opportunities and challenges for individual countries. Differing trade patterns and the effects of unequal power relations within the global trade system can be exemplified by case studies of an AC such as the USA and an LIDC such as Sierra Leone.

Global migration

This section will examine the contemporary patterns of global migration.

Global migration involves dynamic flows of people

The **global migration system** is dynamic and increasingly complex. As globalisation proceeds in the twenty-first century there is greater **connectivity** and growth in the number of places of **origin** and **destination** for international migrants. And there has been growth in the total number of migrants. An estimate in 2015 by UNFPA, the United Nations Population Fund was that 244 million people, 3.3% of the world's population, were living outside their country of origin.

Patterns of international migration can be seen at scales from local cross-border migration to long distance inter-continental movement. Many migrants are **economic migrants**, but there are growing numbers of refugees and asylum seekers. Migration policies, border control, migrant safety and the socio-economic and political impacts of migration have become priority issues for governments and inhabitants of almost all nations.

Current spatial patterns in migrant flows

International migration

The UN definition of a **long-term migrant** is a person who moves to a country other than his or her normal residence for a period of at least one year. A **short-term migrant** is a person who moves for at least three months but less than a year. The overall population change of a country is the result of both natural change and net migration. The term natural change refers to either **natural increase** (crude birth rate > crude death rate) or **natural decrease** (crude birth rate < crude death rate). **Net migration** is the difference (gain or loss) between total numbers of **immigrants** and **emigrants** for a particular country. International migration is therefore one component of population change.

Numbers, composition and direction

The direction or pattern of major international migrant flows in the twenty-first century is shown in Figure 8.

A refugee is a person who has moved outside the country of their nationality or usual domicile because of genuine fear of persecution or death.

An asylum seeker is a person who seeks entry to another country by claiming to be a refugee.

Knowledge check 19

Explain the difference between an economic migrant and a refugee.

Knowledge check 20

What is meant by the term 'net migration gain'?

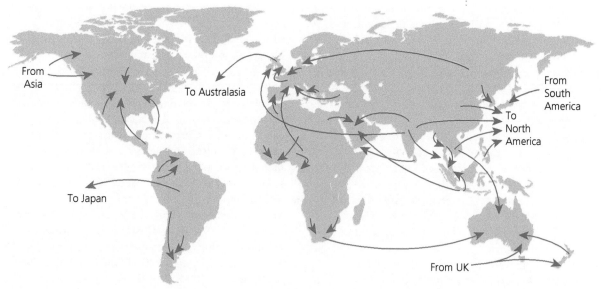

Figure 8 Major global migrant flows in the twenty-first century

Global patterns of migrant flow are complex. As we shall see later, this can be explained by factors such as economic globalisation, the growth in South–South migrant flows and increasing numbers of refugees.

Numbers, composition and directions of migrant flows can be demonstrated by migration statistics for the UK. There is a wide range of countries in all continents with which the UK has connectivity in the global migration system (Table 7). In total by 2014, just over 5 million people born in the UK lived abroad, and there were 7.8 million foreign-born people living in the UK.

The three countries which have supplied the greatest number of immigrants to the UK are India, Poland and Pakistan. The three countries in which most UK emigrants are resident are Australia, USA and Canada.

Exam tip

Appropriate use of key terms such as net migration, refugee or economic migrant is important in all answers. In the shorter data-response sections this allows your responses to be precise and concise.

Table 7 Main migrant connections for the UK, 2014

Country of origin of UK immigrants	Number of immigrants resident in UK (million)	Destination of UK emigrants	Number of UK emigrants resident in the destination (million)
India	0.76	Australia	1.28
Poland	0.66	USA	0.76
Pakistan	0.48	Canada	0.67
Ireland	0.41	Spain	0.38
Germany	0.31	New Zealand	0.31
Bangladesh	0.24	South Africa	0.31
USA	0.22	Ireland	0.25
South Africa	0.21	Germany	0.10
Nigeria	0.18	Channel Islands	0.07
China	0.15	Italy	0.07
Jamaica	0.15	Netherlands	0.05
Kenya	0.15	Switzerland	0.05
France	0.14	Cyprus	0.04
Italy	0.14	Isle of Man	0.04
Sri Lanka	0.13	Poland	0.04

[Source: Pew Research Centre]

The main reasons for UK born people living abroad include: employment opportunities (many UK emigrants are economic migrants in working age groups), retirement (a high percentage of UK emigrants are over 55), family reunification (joining relatives overseas).

Of the foreign-born population living in the UK, Asian countries are the origin of most immigrants. The largest diaspora is by far the Indian population, many living in London. Most migrants who have moved to the UK are economic migrants having secured an employment contract before arrival, but many have arrived seeking work. There is also a high proportion of migrant students in full-time education, and many others seeking family reunification.

Diaspora is the spread of an ethnic or national group from their homeland e.g. Jews from Israel or Kurds from Kurdistan.

Inter-regional migrant flows

Inter-regional migrant flows are exemplified in the last decade by the thousands of people who have risked their lives fleeing conflict in Africa and the Middle East to reach Europe.

The numbers increased significantly in 2015 and this has become a major issue within the EU, especially for Italian and Greek coastguards and Frontex, the EU's border management agency. In addition, a number of organisations such as UNHCR (United Nations High Commissioner for Refugees), many international NGOs (non-government organisations) and national governments have been concerned with migrant welfare, as in some instances entire families, including the very young, attempt dangerous sea crossings at the hands of people traffickers.

Humanitarian organisations not only provide assistance at sea and on the coastline but also along the land routes taken by migrants attempting to travel from the Mediterranean coast northwards into Europe to countries such as Germany and France.

Intra-regional migrant flows

Within the EU, international migration patterns are complex and frequently changing. This is explained by the fact that every EU-28 member state has immigrant populations from other EU states and there is free movement of people across most of the EU's internal borders. The largest recipient of migrants in 2013 was Germany (345,692). Other recipients of large migrant numbers from countries within the EU included the UK (192,495), France (94,393), Spain (85,020) and Italy (75,710).

The size of these flows is explained by political factors such as:

- the Schengen Agreement which, although not applicable to all EU states, allows freedom of movement across internal national borders within most of the EU
- the expansion of the EU, including 13 new states in the twenty-first century, most of which are in Eastern Europe. This has increased the potential number of migrants within the Schengen Area.

The main reasons for these intra-regional migrant flows are economic:

- employment opportunities
- higher wages
- better standard of living
- relative ease of return to the home country after a few years of higher wage earning.

Socio-economic factors add to the reasons for migration within the EU including retirement, joining family members and education.

Friedmann's **core–periphery** model is applicable at continental scale in Europe. The spatial socio-economic **inequalities** which it describes are partly responsible for the direction of migrant flow within the EU. Peripheral areas such as Romania and Bulgaria in Eastern Europe are places of origin of many economic migrants. Their migration is often to destinations in core areas such as Germany and France.

Figure 9 shows Lee's migration model which is a useful framework for understanding causes of international migration and factors that influence migrants between their countries of origin and destination.

Push factors are negative factors which apply in a migrant's current location such as the effects of conflict. **Pull factors** are perceived advantages of a potential destination which attract migrants such as employment opportunities. **Intervening obstacles** can have an influence at any point from origin to destination and include physical, social, economic and political factors including costs of travel, crossing a sea, or use of language.

Exam tip

You should be prepared to identify features of spatial patterns shown on different types of statistical maps such as choropleths or flow line maps. This could involve patterns of inter- and/or intra-regional migrations.

Exam tip

Ability to write clear definitions of geographical terms such as 'economic interdependence', 'migrant remittances' or 'inequalities' can be helpful in essays. This demonstrates your understanding of key concepts which are important in the discussion.

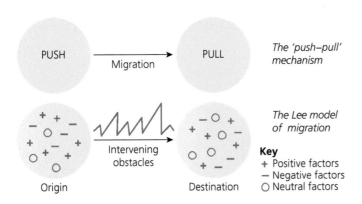

Figure 9 The push-pull mechanism and the Lee model of migration

International migration and socio-economic development

The relationship between current patterns of international migration and socio-economic development

The relationship between patterns of international migration and development can be demonstrated by statistical evidence such as migrant remittances and HDI. **Migrant remittances** are private funds sent by economic migrants usually back to their families who remain in the country of origin. These remittances are very important in the development process and are of particular significance to the poorest countries. The **Human Development Index (HDI)** is a composite index which is the UN's measure of development based on economic and social indicators.

Globally there is a relatively strong inverse relationship between receipt of migrant remittances and HDI. ACs tend to receive amounts of migrant remittances which are only a low percentage of **Gross Domestic Product (GDP)** whereas for LIDCs and EDCs migrant remittances are a much higher percentage of GDP. For example, they constitute around 20% of GDP in LIDCs such as Haiti, The Gambia and Liberia. These are all countries with relatively low HDI. Migrant remittances to ACs such as the USA and UK, with high HDI, represent only about 0.1% of GDP or less.

There is increasing recognition that migration can be a factor in the promotion of international development. The extent to which there is a causal relationship between remittances and socio-economic development depends on how migrant remittances are used. The money often has the effect of lifting families back home above poverty levels. It may be spent on improving housing and quality of life or invested in small-scale local business enterprises such as retailing. It can help to stimulate the economic multiplier effect by creating local employment opportunities and it may enable easier access to transport, health and education services.

Knowledge check 21

Outline the main difference between push and pull factors in the context of migration.

Exam tip

Ability to describe the relationship between two sets of data with reference to a scatter graph, a table or other statistical diagrams is useful. You should be able to explain the relationship and any anomalies, such as that between migrant remittances and HDI.

Knowledge check 22

Explain the importance of migrant remittances in the process of development.

Global migration can promote stability, growth and development

Global migration can promote stability, growth and development within and between countries.

The importance of migration as a key factor in the development process is recognised in the United Nations Development Programme (UNDP). International migration has positive impacts on economic and political stability, economic growth and socio-economic development in countries of origin and destination.

Stability

Migration can promote stability in the following ways.

- Migrant remittances as a source of foreign exchange can contribute to the economic stability of a recipient country.
- Returning migrants, having acquired new ideas and values including democracy and equality, can contribute to peacebuilding and conflict resolution.
- Where there is an ageing population, youthful migrant working populations contribute to a more balanced age structure and population growth.

Economic growth

Migration can promote economic growth in the following ways.

- The GDP and tax base of the host nation can be boosted by working migrants.
- Migrants are consumers in a host country and stimulate local economies, even opening up new markets in demands for food, clothing, music, etc.
- Migrants can fill skills gaps and shortages in the labour market of a host country at local and national scales.
- Migrant remittances can supplement household income, increase consumer spending, provide funds for local investment and stimulate local **multiplier effects** in the country of origin of the migrants.

Development

Migration can also promote development.

- Skills and knowledge acquired by returning migrants can be of benefit to families and local communities and also may be of national value in countries of origin.
- Migrants can create networks which ease flows of skills, financial resources, values and ideas through their links to diaspora associations, including professional, business, social and religious networks.
- UN 'migration and development' projects between partner countries involve families, local authorities, and public and private service providers in effective 'bottom-up' approaches to development.

Flows of people, money, ideas and technology are linked to these positive effects of international migration, for example as follows.

- Migrant remittances are very significant transfers of money worldwide. According to the World Bank they are worth hundreds of billions of dollars each year. These flows have been made easier and more reliable by cash transfer programmes and use of smart phone technology.

Exam tip

Responses to questions on the impact of international migration should consider different perspectives. These impacts may be viewed as economic, social, political or environmental; positive/advantages or negative/ disadvantages; short- or long-term; and local scale through to global scale.

- International migration leads to spatial diffusion of ideas, information, and values. These are transmitted back to country of origin by migrants and include **social remittances** such as flows of ideas about family size, education and marriage.
- Technology is increasingly important in the work of international humanitarian organisations. Examples include: biometric technology using iris scans to register migrants in refugee camps and the use of crowd-sourced data from text messages, emails and tweets in 'crisis mapping'.

Global migration causes inequalities, conflicts and injustices

International migration can cause inequalities, conflict and injustices for people and places, in countries of origin and destination.

Inequalities

Global migration causes inequalities in the following ways.

- Countries of origin lose a proportion of the young, vibrant and fittest element of the labour force. This may contribute to a downward economic spiral at local, regional and national scales.
- Often it is the better educated that migrate. This represents a 'brain drain' and loss of human resources in the country of origin.
- The demographic selectiveness of international migration causes redistribution of the population of reproductive age. This influences crude birth rates in countries of origin (decline) and destination (growth).
- Migrant remittances can increase inequality between families who receive them and those who do not within a local area.

Conflicts

Global migration causes conflicts in the following ways.

- Social conflict can develop between host communities and 'newcomers'. People of a particular culture or ethnic origin may find difficulty integrating, perhaps because of language.
- Immigrant populations, especially if concentrated in specific areas, can place pressure on service provision such as education, health, and housing, in the host country,
- International borders can be areas of conflict between border control authorities and human traffickers and illegal migrants.

Injustices

Global migration causes injustices in the following ways.

- Migrants are vulnerable to violation of their human rights as a result of forced labour, exploitation of women and children and human trafficking.
- Treatment of asylum seekers can include being held in detention centres, not being allowed to work, being supported on meagre financial resources for food, sanitation and clothing for the duration of their application.
- The plight of refugees in terms of shelter, food, water, medicines and safety, including the possibility of return to their country of origin where risks are high.

Knowledge check 23

What is meant by the term 'diaspora'?

Exam tip

Essay questions requiring evaluation of migration-related issues should have a discursive response that includes a range of arguments, a balanced assessment of the issues, and a conclusion consistent with that balance.

Unequal flows of people, money, ideas and technology are linked to inequalities, conflicts and injustices, for example as follows.

- Migrant flows are spatially uneven and irregular over time. South–North and South–South corridors are the dominant directions of flow within the global migration system. Transfers of financial and social remittances are closely related to these patterns of inequality and tend to flow in the opposite direction to the migration within these corridors — that is, back to the country of migrant origin.
- Access to technology is also spatially uneven. At the global scale internet access, for example, is least available in countries of lower socio-economic status such as in sub-Saharan Africa. This can restrict the work of NGOs and other relief organisations who rely on effective communication and transfer of information in crisis areas where refugees experience conflict and injustices.

Why has migration become increasingly complex?

Global migration patterns are influenced by a multitude of interrelated factors. Changes in the twenty-first century have increased the complexity of global migration.

Factors influencing global migration patterns

Economic globalisation

New countries of origin and new countries of destination have emerged as the effects of **economic globalisation** have spread and intensified in the twenty-first century. The growing complexity of the global migration system and the growing **interdependence** of countries and their economies can be exemplified by flows of economic migrants at different scales.

Inter-regional

At this scale migration includes the following.

- Migration of highly skilled workers and graduates in science, mathematics and technology from EDCs such as China, India and Brazil to ACs such as the USA and Canada, attracted by high salaries.
- Migration of workers from India, Bangladesh, Pakistan, Egypt, the Philippines and Indonesia to the oil-producing Gulf States and Saudi Arabia, attracted by relatively high wages and the opportunity to send remittances home.

Intra-regional

At this scale migration includes the following.

- Rapid increase of low-skilled international migrant stock within ASEAN from the poorer countries of Myanmar, Lao PDR and Cambodia to the faster growing economies of Singapore, Malaysia and Thailand. Thailand's economic growth has created demand for unskilled labour, for example in construction.
- Increased migrant streams within South America to the 'southern cone' of Argentina, Chile, Paraguay and Uruguay, driven by wage differentials and labour opportunities.
- Return migrations in the EU from taking low-skilled jobs in richer countries such as Germany, UK and Italy back to countries such as Romania, Lithuania, Latvia, Portugal, Poland and Estonia, often returning to more prestigious positions.

Knowledge check 24

What is the difference between 'origin' and 'destination' in the context of international migration?

Exam tip

Be prepared to identify patterns and trends over time from statistical tables and graphs such as line or bar charts. This could include data on changes in total number or demographic composition of international migrant flows, flows in bilateral migrant corridors, or changes in indices of development such as HDI for particular countries.

International migrant stock is the number of people born in a country other than that in which they live, including refugees.

Internal

Rural–urban migration is a major element of the global migration system. This has been reinforced in the twenty-first century by the effects of FDI in or near large urban centres in EDCs such as India, China, Mexico and Brazil in particular. This has created a range of employment and social opportunities.

Knowledge check 25

What do you understand by 'internal migration'?

High concentrations of young workers and female migrants

Young workers

Most economic migrants, seeking greater employment opportunities, higher wages and the possibility of sending remittances, are young. This is clearly demonstrated in Figure 10 by the 25–39 age group of migrant populations in Asia, 2013.

An example is the labour-driven migration of young workers to the oil-producing countries of UAE, Qatar and Saudi Arabia. In 2013 the population of UAE alone included 7.83 million foreign-born people, the majority of whom were low-skilled, young males, working in construction.

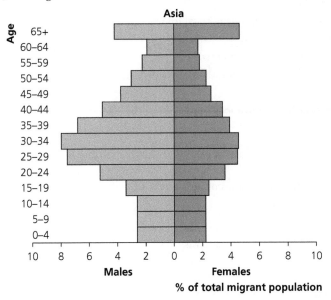

Figure 10 The age–sex distribution of international migrants in Asia, 2013

Female migrants

There has been an increase in women and girls in the global migration system. In some regions such as Europe, Latin America and the Caribbean, North America and Oceania, the number of female international migrants has now exceeded that of men. This is the result of increasing female independence and status, greater freedom and the growing importance of women as main income earners. This is not the case for Africa and Asia (Figure 10) where factors such as regulations governing female admission and departure from some countries and the status of women are restrictive.

A growing trend is the migration of highly skilled, graduate women especially to countries where there is less discrimination in the labour market and where in general women's rights are better respected.

Exam tip

When giving explanations of spatial patterns or trends over time for migration data, refer to the influence of specific types of physical, economic, social and political factors. Also demonstrate your understanding of how these factors are interrelated or apply in combination.

UNHCR, United Nations Population Fund (UNFPA), International Labour Organization (ILO) and International Organization for Migration (IOM) are international organisations which are involved in the governance of female migration and the wellbeing of female migrants.

Increased flows in South–South corridors

Until 2013 the largest international migrant flows were from the poorer, less developed countries of the South to the wealthier more developed countries of the North. But now the international migrant stock of South–South corridors outnumbers that of the South–North corridors.

Recent increase in South–South flows can be explained by:

- restrictive administrative barriers for migrants from the South attempting to enter the North. Often in response they redirect their migration to another South country
- the number of fast growing economies in the South which offer employment opportunities and are increasingly accessible
- increased awareness of opportunities in the South resulting from improved communications and development of social and business networks
- the preventative costs of moving to more distant richer countries
- increased ease of migration within trading blocs in the South such as ASEAN and MERCOSUR
- improvements in ease and reliability of sending remittances between South countries
- increase in numbers of refugees fleeing conflict or persecution in South countries.

Developing countries (LIDCs and EDCs) are the destination of 86% of the world's refugees, such as flows from Afghanistan to Pakistan.

Examples of South–South corridors of international migration include flows between Burkina Faso and Ivory Coast and between Myanmar and Thailand.

Increased numbers of refugees

According to UNHCR the number of refugees worldwide increased from 15.7 million in 2012 to 19.5 million in 2014. By 2015 Syria had become the top source of refugees, overtaking Afghanistan which had held this position for three decades. Countries with the highest ratios of refugees to total population in 2014 were Lebanon (257/1000) and Jordan (114/1000), which both border Syria.

The main reasons for the large number of refugees globally include:

- the effects of conflict, including personal safety, loss of homes, restricted access to services, and damage to infrastructure including communications
- political persecution, discrimination and violation of human rights
- economic hardship and persecution as a result of forced labour and modern slavery
- the impacts of natural hazards.

During 2014, conflict and persecution forced 42,500 people per day to leave their homes and move either within the borders of their country as internally displaced persons (IDPs) or to other countries as refugees. Approximately half of all refugees were under 18 years old.

In 2014, 1.66 million people had submitted asylum applications. The largest recipient countries of these applications were the Russian Federation, Germany and USA.

Changes in national immigration and emigration policies

Migration policies of national governments are designed to meet the specific social, economic and political needs of those countries. These requirements differ widely, therefore policies on immigration and emigration vary from time to time and from place to place. A useful distinction is that some ACs, such as Australia and Canada, have at times encouraged immigration using a points-based system to satisfy labour shortages, and some LIDCs, such as Pakistan, actively encourage emigration for the socio-economic benefits it can bring to their overall development through financial and social remittances.

Immigration policy

The immigration policies of some ACs, such as Canada, are based on a points system. Canada has shortages in engineers, IT specialists and health care workers. The aim is to fast-track young highly skilled workers in order to address shortages in these sectors if the positions cannot be filled by Canadian citizens.

The USA immigration policy has an annual worldwide limit of 675,000 permanent immigrants. The policy allows immigration for: reunification of families, economic migrants with skills valuable to the US economy, protection of refugees, promoting diversity to maintain a balance of immigrants from different countries and humanitarian relief.

Emigration policy

The ILO, a specialised agency of the UN, works with governments in drafting migration policies. Their aim is to promote international labour migration, to protect migrant workers and to support diaspora communities. In the case of LIDCs and EDCs, the ILO helps to ensure that:

- rights of workers abroad and their basic human rights are protected
- there are training schemes for potential migrants in preparation for working abroad
- the value of the manpower of the developing country (LIDC or EDC) is promoted in potential recipient countries
- female participation in overseas employment is encouraged — currently only 0.12% of labour migration from Pakistan, for example, is female
- there is support for the social networks and associations of the diaspora
- advice is provided on effective use of economic and social remittances of returning migrants to assist in the development process.

Corridors of bilateral flows

Distinct corridors of bilateral flows have developed. Bilateral migration corridors involve migrant flows between two countries. The flows vary in number of

Internally displaced persons (IDPs) are people who have been forced to move within their own country as a result of conflict or environmental disaster.

Knowledge check 26

What is an asylum seeker?

migrants and in their demographic composition. The largest bilateral corridor is between Mexico and the USA. The average annual flow of Mexicans to the USA between 2000 and 2010 was 250,000, but the number has declined in recent years (Figure 11).

New corridors with significant numbers have been recorded in the last decade. These add to the complexity of international migration. Examples include flows of economic migrants within the ASEAN Economic Community (AEC) and migrations between countries of Eastern Africa which include many refugees.

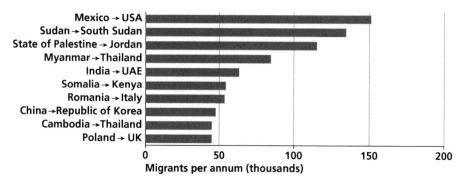

Figure 11 Bilateral migration corridors with the largest number of international migrants per annum, 2010–13

Bilateral corridors develop as a result of economic, social and political factors such as:
- costs of travel
- ease of access and communications
- efficiency, reliability and costs of sending remittances
- employment opportunities and wage differentials
- establishment of diaspora communities and networks
- effects of conflict and persecution
- policies of national governments and economic/political unions
- legacy of former colonial influence such as language.

Corridors of migrant flows create interdependence between countries

International migration is an important part of the globalisation process and it contributes significantly to social and economic interdependence between countries. For example, socio-economic interdependence occurs where economic migrants provide labour in a country and on return bring newly acquired skills, ideas and values to their home country. Increasingly countries linked through migration are becoming interdependent in trade, and in political and environmental relationships.

This aspect of globalisation has created advantages and opportunities for many people and places, but others have been marginalised and experience the effects of inequality and injustice.

To illustrate this topic and show full understanding of the term interdependence, the specification requires a case study of an EDC such as Brazil or India.

Exam tip

When using case material to support an argument or a particular issue, refer to place-specific details as well as statistical evidence. It is helpful to learn place names, and key migration and development statistics for your chosen EDC, AC and LIDC examples.

Case study of one EDC

For your chosen case study, you should be able to illustrate the following.
- Current spatial patterns of immigration and emigration, including statistics for the main flows.
- Changes in immigration and emigration over time. This could include both recent changes, such as the effects of trading bloc membership on flows, and also historic flows, perhaps linked to the colonial period, and their legacy.
- Examples of economic, social, political and environmental interdependence with countries connected to the EDC through migration. This could include relationships with two or three bilateral migration partners. For each you

should investigate: economic interdependence such as employment opportunities and financial remittances, socio-economic interdependence such as social remittances and development of diaspora networks, political interdependence, for example in negotiating trade agreements and security, and environmental interdependence, for example cooperation between governments over issues such as atmospheric pollution or protection of natural ecosystems.
- The impact of migration on the EDC's economic development, political stability and social equality. This could include both opportunities and challenges at different scales.

The issues associated with unequal flows of global migration

Global migration creates opportunities and challenges

This part of the specification requires two case studies to illustrate how, within the global migration system, advanced economies, such as the USA, have the power to influence and drive change, and peripheral economies, such as Lao PDR, Haiti or Ethiopia, have limited power and influence. You should also investigate the impact of these unequal power relations on people and places in terms of opportunities and challenges created by migration.

Knowledge check 27

In terms of international migration, explain what is meant by socio-economic interdependence.

Case study of one AC

For your chosen AC case study, you should be able to illustrate the following.
- Patterns of immigration and emigration. This should include the main countries of origin of immigrants, reasons for the attractiveness of the AC for immigrant populations, the main destination countries for emigrants from the AC and reasons for emigration. These patterns of migration should be supported by statistical evidence.
- Details of the AC's migration policies and their purpose.
- An understanding of the interdependence of the AC with countries linked to it by international migration. This could include the economic, social, political and environmental relationships developed by the AC in its bilateral migration corridors.

- Opportunities and challenges created within the AC as a result of international migration. Opportunities could include addressing labour shortages in both low-skilled and highly skilled jobs in different sectors of industry, stimulating local multiplier effects and demographic effects, such as the impact of increased crude birth rate on the population structure of the AC. Challenges could include the problems for border control in dealing with unauthorised immigrants, people trafficking, contraband and security, the integration of immigrant populations into the society of the host AC, and the supply of services and resources where immigrant populations are concentrated.

Case study of one LIDC

For your chosen LIDC case study, you should be able to illustrate the following.

- Patterns of emigration and immigration. This should include the main destination countries for emigrants from the LIDC, reasons for emigration, the main countries of origin for immigrants to the LIDC, and reasons for immigration. These patterns of migration should be supported by statistical evidence.
- Details of the LIDC's migration policies and their purpose, plus any additional laws on migration to which the LIDC is subject as a member of a trading bloc.
- An understanding of the interdependence of the LIDC with countries linked to it by international migration. This could include the economic, social, political and environmental relationships developed by the LIDC in its bilateral migration corridors.
- Opportunities and challenges created within the LIDC as a result of international migration. Opportunities could include stimulation of political and economic cooperation with other countries, especially where bilateral links are strengthening, the specific benefits of migrant remittances at different scales, and increasing political stability. Challenges could include the effects of the 'brain drain' including loss of low-skilled and highly skilled labour, and problems of exploitation of workers by human trafficking, forced labour and exploitation.

Exam tip

You should be able to explain the unequal power relations within the global migration system and the issues arising from these inequalities for one AC and one LIDC.

Knowledge check 28

Identify three challenges for LIDCs resulting from international migration.

Summary

- The global migration system is dynamic. In the twenty-first century there has been significant growth in the number of international migrants.
- Spatial patterns of migrant flows are complex and they are influenced by economic, political, social and environmental factors.
- Spatial patterns of migrant flows are closely related to global patterns of socio-economic development.
- International migration can have the effect of promoting stability, economic growth and socio-economic development. It can also be a cause of inequalities, conflict and injustice.
- The global migration system not only incorporates flows of people of varied demographic composition but also flows of money, ideas and technology. These have an impact on international migration and development.
- The global migration system has become increasingly complex. This is the result of changes in the twenty-first century which include: the emergence of new source areas and host destinations, a high concentration of young workers and female migrant stock, flows in the South–South corridors becoming equal in magnitude to those in South–North corridors, the effects of conflict and persecution in increasing the number of refugees, changes in national immigration and emigration policies, and the development of distinct corridors of bilateral migrant flows.
- There has been increasing interdependence between countries created in the corridors of migrant flow. Economic interdependence is extended to social, political and environmental interdependence as the relationships between countries develop as part of the globalisation process. This can be exemplified by a case study of an EDC such as Brazil.
- The effects of unequal power relations on international migration within the global migration system create both opportunities and challenges for individual countries. These can be exemplified by case studies of an AC such as the USA and an LIDC such as Lao PDR.

Human rights

What is meant by human rights?

Global variation in human rights norms

Human rights are the basic rights and freedoms to which all human beings are entitled. They are applicable at all times and in all places and they protect everyone equally, without discrimination.

The **Universal Declaration of Human Rights (UDHR)**, adopted by the United Nations General Assembly in 1948, contains 30 articles which define the various human rights.

Examples include:
- No one shall be subjected to torture or to cruel inhuman or degrading treatment or punishment (Article 5).
- No one shall be subjected to arbitrary arrest, detention or exile (Article 9).

Not all human rights and freedoms set out in the UDHR have been adhered to uniformly throughout the world. Violations have occurred at all scales in every continent including both advanced (ACs) and developing countries (EDCs and LIDCs).

Geographical patterns of socio-economic inequality are closely associated with respect for human rights. Many development programmes, including the UN's Millennium Development Goals (MDGs) and the post-2015 Sustainable Development Goals (SDGs) are human rights led.

Understanding the terms 'norms', 'intervention' and 'geopolitics'

Human rights norms

Statements set out in the UDHR are generally accepted as **international human rights norms**. These were drawn up on the basis of established customs and ways of living from all countries, religions and philosophies across the world. These norms are the moral principles that underpin universally accepted standards of human behaviour. They are protected by international law.

International human rights law sets out obligations of state governments. It is the duty of states to respect, protect and fulfil international human rights. Governments that sign **international treaties** must put in practice domestic measures and legislation compatible with that treaty.

There are growing numbers of human rights norms, laws and treaties. One of the most widely ratified is the **United Nations Convention on the Rights of the Child (UNCRC)**. Despite this there is still significant global variation in deaths of young children (Figure 12).

Sustainable Development Goals (SDGs): the 17 SDGs are included in the 2030 Agenda for Sustainable Development to promote universal development, end poverty, fight inequality and injustice and tackle climate change.

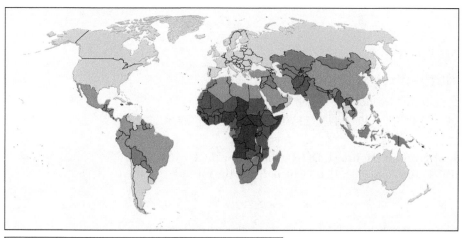

Key
Number of deaths of infants under age 1 per 1000 live births.
▮ 75+ ▮ 50–74 ▮ 25–49 ▮ 15–24 ▯ <15

Figure 12 Global pattern of infant mortality rates, 2013

Most of these infant deaths could be prevented. The UN view is that if a country is not doing what it can to prevent these deaths it is not meeting its legal and moral obligations. Therefore **infant mortality** is not just a health matter but a human rights concern.

Intervention

Intervention includes:

- the use of military force by a state or group of states in a foreign territory to end gross violation of fundamental human rights of its citizens
- economic sanctions
- international criminal prosecution of individuals responsible for human rights abuses
- the work of civil society, such as NGOs, private enterprises and individual human rights activists often in local communities.

Use of force can be authorised legally only by the **United Nations Security Council**. The entire process of military intervention is controversial. It can bring benefits in both the short and long term, but it can also have unintended negative impacts on people and the places in which they live, including further injustices and human rights violations.

UN involvement includes its peacekeeping, political and peacebuilding missions. UN workers and troops are drawn from a wide range of member states to protect citizens from human rights abuse, to protect and promote human rights, to monitor the situation, to empower people to assert their human rights, and help governments to implement them and strengthen rule of law. The UN also coordinates the work of other agencies and organisations, for example in an area of conflict. These could include the North Atlantic Treaty Organisation (NATO), the Organisation for Security and Cooperation in Europe(OSCE), the ASEAN Intergovernmental Commission on Human Rights, NGOs such as the International Committee of the Red Cross (ICRC), public partnerships such as the Global Alliance for Vaccines and Immunisation (GAVI), and the national governments of the area concerned.

> **Exam tip**
> Correct use of terminology will help your answers to be more concise and authoritative. You should be able to demonstrate clear understanding of terms such as 'human rights norms', 'intervention', 'geopolitics' and 'global governance'.

> **Knowledge check 29**
> What do you understand by the term 'human rights norms'?

> **Civil society** is NGOs and other organisations that are independent of governments, working voluntarily, either individually or collectively, in support of citizens and communities throughout the world.

The term global governance of human rights therefore involves a complex and multifaceted set of processes including: direct physical intervention, the application of human rights norms, laws and treaties, and the work of civil society. The interaction and cooperation of all involved at different scales is essential for successful outcome.

Geopolitics

The term **geopolitics** refers to the global balance of political power and international relations. Contemporary patterns of political power are closely related to economic power and have an uneven spatial distribution:

- Globally the USA is dominant militarily and politically.
- Inequalities exist between powerful ACs, increasingly influential EDCs and LIDC peripheral economies.
- Supra-national political and economic organisations such as the UN, EU and OPEC exert strong geopolitical influence.
- Trans-state MNCs have considerable influence on global trade and the countries in which they invest.

The term **geopolitics of intervention** involves an understanding of the political composition of groups of countries and organisations involved in intervention, the nature of the intervention itself, the reasons why intervention is deemed necessary, the characteristics of the country, government and people affected, and possible political, socio-economic and environmental effects of the intervention.

Patterns of human rights violations

Patterns of human rights violations are influenced by a range of factors.

Current spatial patterns of human rights issues

'Everyone has the right to life, liberty and security.' (Article 3, UDHR). Forced labour, maternal mortality rates and capital punishment are all connected to this most basic of human rights. But global patterns reveal significant variation in their prevalence.

Forced labour

Forced labour is described by the ILO as 'situations in which persons are coerced to work through the use of violence or intimidation or by more subtle means such as accumulation of debt, retention of identity papers or threats of denunciation to immigration authorities'. This is a significant element of modern slavery.

There are estimated to be 21 million victims of forced labour worldwide. These include:

- children denied education because they are forced to work
- men unable to leave work because of debts owed to recruitment agents
- women and girls exploited as unpaid, abused domestic workers.

The global pattern is uneven but no world region is unaffected. Estimates for 2012 show that highest incidence was in southeast Asia, with 11.7 million. The figure for Africa was 3.7 million and overall in North America, Europe and Australia and New Zealand it was 1.5 million.

Factors that influence global variation of forced labour

Economic, political, social and environmental factors that contribute to vulnerability of forced labour are outlined in Figure 13.

Global governance is the intervention by the global community, attempting to regulate issues, such as human rights, sovereignty and territorial integrity.

Knowledge check 30

What is meant by intervention by the global community in human rights issues?
....

Exam tip

Inequalities in spatial patterns of human rights violations are explained by a range of economic, social, political and environmental factors. You should be able to demonstrate how these factors are important, often in combination.
....

<table>
<tr><td>

ECONOMIC

- Poverty
- Lack of economic opportunities and unemployment
- Low wages
- Subsistence farming
- Migration seeking work

</td><td>

POLITICAL

- Political instability
- Conflict
- Breakdown of rule of law
- Corruption
- State sponsorship of modern slavery e.g. cotton harvest Uzbekistan
- High levels of discrimination and prejudice

</td></tr>
<tr><td>

SOCIAL

- Gender inequality
- Age, especially children
- Entire families enslaved through bonded labour e.g. construction, agriculture, brick making, garment factories in India and Pakistan
- Women and children trafficking for sexual exploitation e.g. through organised crime in Europe from Nigeria
- Indigenous people

</td><td>

ENVIRONMENTAL

- Escaping climate related disasters including food and water shortages
- Hazardous working conditions in open mines

</td></tr>
</table>

Figure 13 Factors that contribute to forced labour

Maternal mortality rate

The **maternal mortality rate (MMR)** is the annual number of deaths of women while pregnant, or within 42 days of termination of pregnancy, from any cause related to or aggravated by the pregnancy or its management, per 100,000 live births (WHO).

In 2013, globally 289,000 women died during and following pregnancy and childbirth. Most of these deaths were in developing countries (LIDCs and EDCs) with the highest rates in sub-Saharan Africa and the lowest in Europe and North America. For example, the figures per 100,000 live births for Sierra Leone and Italy were 1,100 and 4 respectively.

Factors that influence global variation of MMR

Global inequalities in MMR are explained by variations in: access to treatment, especially emergency care, the quality of medical services especially skilled attendants at birth, the level of political commitment and government investment, the availability of information and access to education, cultural barriers such as discrimination, and poverty.

Many of the deaths are preventable. Therefore this is not just a matter of development but of human rights. One international human rights treaty which affords legal protection is the Convention on the Elimination of all forms of Discrimination Against Women (CEDAW).

Capital punishment

Capital punishment is a fundamental breach of human rights. The UN General Assembly has called for an end to the death penalty. Nevertheless there were over 600 executions globally in 2014. There are significant global inequalities in the spatial pattern of executions and death sentences. High numbers of executions have been reported in China and in some countries of the Middle East. In Europe, only Belarus retains the death penalty.

Factors that influence global variation of capital punishment

The inequalities are explained by: differences between countries in the range and type of crimes for which the death penalty is imposed, variations in its legality under

Knowledge check 31

What is meant by 'forced labour'?

Knowledge check 32

At the global scale, what is the relationship between spatial patterns of maternal mortality rates and levels of development?

national law, the increase in the number of countries in which it has been abolished, its reinstatement in some countries in response to threats to state security and public safety posed by terrorism, and the number of commutations and pardons.

Variations in women's rights

This section examines the geography of gender inequality. **Gender inequality** refers to situations in which women and men do not enjoy the same rights and opportunities specifically because they are a woman or a man. The geography of gender inequality is complex and contested, but the extent of unequal treatment based on gender can be measured by a number of indices. Examples include labour force participation, healthy life expectancy, enrolment in secondary education, and participation of women in parliament. These show that women are not treated equally to men. This is a serious obstacle to development.

The Global Gender Gap Index (GGGI), devised by the World Economic Forum (WEF), is a measure of gender inequalities throughout the world. Others include the Gender Inequality Index (GII) used in the United Nations Development Programme (UNDP). Gender inequality statistics are closely related to development statistics such as the Human Development Index (HDI) — see Table 8.

Table 8 GGGI and HDI statistics for selected countries, 2014

Country	GGGI	HDI
Iceland	0.859	0.895
Finland	0.845	0.879
Norway	0.837	0.944
Sweden	0.816	0.898
Denmark	0.802	0.900
Switzerland	0.779	0.917
UK	0.738	0.892
Brazil	0.694	0.744
Russia	0.692	0.778
China	0.683	0.719
India	0.645	0.586
Iran	0.581	0.749
Mali	0.577	0.407
Syria	0.577	0.658
Chad	0.576	0.372
Pakistan	0.552	0.537
Yemen	0.514	0.500

Table 8 shows that Nordic countries in particular have high values for both indices, whereas some countries of the Middle East and North Africa are low in both.

Despite improvements in protecting and promoting human rights in the twenty-first century, there are remaining challenges of gender-based discrimination and exploitation of women and girls. These inequalities are explained by factors such as forced marriage, trafficking into forced labour, access to education and health care, employment opportunities including political participation, wage equality, violence,

Exam tip

In the analysis of statistical maps and diagrams, you should be able to describe and explain the patterns and trends shown, and evaluate the effectiveness of the technique used to represent the data.

and access to reproductive health services. These are reinforced in some countries by deeply rooted patriarchal norms.

Improving the lives of women and female empowerment are the aims of various development programmes of international organisations such as the UN and civil society organisations (CSOs), for example the International Center for Research on Women (ICRW), a US NGO. The UN aims to reinforce norms which outlaw gender discrimination, through international treaties such as The Convention on the Elimination of all Forms of Discrimination Against Women (CEFAW) and the post-2015 Sustainable Development Goals. Many NGOs are working to resolve the problems at local scale within communities, including in their education programmes about the roles of men and boys as well as women and girls.

Economic, political and social factors explain patterns of gender inequality

Economic, political and social factors, such as educational opportunity, access to reproductive health services and employment opportunity can explain variation in the patterns of gender inequality.

Educational opportunity

Gender inequality in education favours males in many LIDCs and EDCs. Although female enrolment has improved especially in primary schools under the UN's Millennium Development Goals, many girls still experience exclusion in education. This is especially the case in the poorest countries of sub-Saharan Africa and in particular amongst the rural poor.

Female education is important in empowering women and achieving gender equality. It helps incorporate women into the labour force and it has a positive impact on **total fertility rates**, population growth rates, **infant mortality rates**, family health, child nutrition and poverty reduction in families.

The UN has set up the Girls Education Initiative, and Sustainable Development Goal 4 focuses on ensuring inclusive and equitable quality education, promoting life-long learning opportunities for all. Also NGOs and some MNCs, as part of their corporate social responsibility, are involved in education partnerships in poor countries.

But there are still many problems to overcome, such as patriarchal societies, household obligations for girls, prohibitive costs in continuing secondary education, problems for girls at school such as negative classroom environments and inadequate sanitation, and relatively few female teachers. In addition girls' education can be restricted by exploitation for child labour, child marriage, early pregnancy, and differing levels of support for education by different religions.

Access to reproductive health services

Female reproductive health rights are violated when women and girls are denied access to health care services. Those living in poor communities in LIDCs and EDCs are most at risk, especially where women are economically and socially disadvantaged.

Figure 14 outlines factors affecting female reproductive health in developing countries. These are mainly social/cultural factors which are discriminatory and closely linked to human rights.

Exam tip

You should understand that the relationship between patterns of gender inequality and levels of development are explained by economic, social, political and environmental factors and that these factors are interrelated.

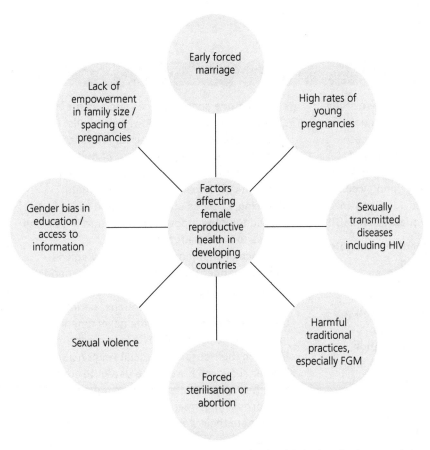

Figure 14 Factors affecting female reproductive health in developing countries

The **Office of the UN High Commissioner for Human Rights (OHCHR)** and NGOs such as Amref Health Africa and Womankind are examples of international organisations involved in resolving these issues working closely 'in the field' with local communities.

Employment opportunity

There is significant global variation in equality of access to employment opportunities for women and men. The Labour Force Participation Rate index used in the UNDP shows very limited female access to the labour market in relation to men in North Africa and the Middle East and in Afghanistan. There is a higher proportion of female participation in countries with high HDI such as the USA and UK. India has a relatively low ratio compared with other rapidly emerging economies.

The inequalities are explained by a range of interrelated social, economic and political factors, including:

- social norms in countries where responsibility for securing household income is attributed to men and where women are expected to devote time to unpaid domestic care
- levels of government and company support for child care
- the degree to which equal opportunities are safeguarded by law
- social acceptance of women as contributors to household income
- gender-based norms that shape educational and job decisions of women and men

Knowledge check 33

Suggest reasons why female empowerment is important in the development process.

Exam tip

You should be prepared to evaluate the relative importance of the different factors that contribute to violation of the various human rights.

Knowledge check 34

What is meant by the term 'global governance'?

Exam tip

Case studies are important in the more extended responses. The application of statistical and place-specific detail of your chosen areas or countries is useful in supporting discussion and in reinforcing arguments.

- levels of discrimination by employers
- the sectoral structure of the labour market.

Case study of women's rights

This part of the specification requires a case study of any one country to illustrate gender inequality issues apparent in that country, consequences of gender inequality on society and evidence of changing norms and strategies to address gender inequality issues. Possible choice of country could include India or Afghanistan ranked 135th and 147th respectively in the UN Gender Inequality Index (GII).

For your chosen case study you should investigate and be able to illustrate the following.
- Gender inequality issues. These could include issues such as violence against women, modern slavery, patriarchal systems of inheritance and property ownership, employment opportunities, discrimination in the workplace, political participation such as female representation in parliament, and access to education and health care.
- Consequences of gender inequality on society. This will depend on the chosen example, but could include consequences such as the effects of being unable to meet dowry demands, the effects of not agreeing to arranged marriage, domestic abuse, high incidence of maternal and infant mortality and morbidity, sex-selective abortions perhaps in the desire for male offspring, coerced sterilisation and the effects of limited female empowerment on family size and access to the workforce.
- Evidence of changing norms or strategies to address gender inequality. This also depends on the chosen country and could include ratification of various treaties or conventions, national and state laws passed by the government, interventions by specific NGOs and adherence of MNCs to their corporate social responsibilities. In your research it would be useful to consider how the various organisations cooperate in addressing gender inequality issues and coordinate their work at different scales.

Strategies for global governance of human rights

Strategies for the global governance of human rights include:
- the work of NGOs, private organisations and human rights activists, including attempts to change norms
- the influence of MNCs in meeting their Corporate Social Responsibility (CSR)
- creation of new laws and attempts to strengthen the rule of law
- the use of international legal mechanisms such as the International Criminal Court (ICC)
- ratification of treaties established by supra-national organisations
- the work of various UN agencies
- the role of the UN in its peace-keeping missions
- military intervention and relief assistance.

Human rights violations can be a cause and consequence of conflict.

Human rights violations can cause conflict

Often there is no single cause of conflict but a combination of contributory factors, such as:
- denial of basic human needs, including food, housing and access to education, over long periods
- discrimination and denial of freedom in undemocratic societies

Knowledge check 35

Outline the main ways in which the United Nations is involved in resolving human rights conflicts.

- unequal or unjust treatment and exclusion from decision making by unrepresentative governments
- lack of respect for ethnic identity or gender discrimination by oppressive governments
- genocide and torture.

Human rights violations can be a consequence of conflict

Conflict may lead to further human rights violation or its intensification, such as:

- increase in the number of deaths of military personnel and civilians
- damage to property in local neighbourhoods and to infrastructure such as transport systems, communications, hospitals and schools
- restricted access to food and water supplies and other services
- displacement of population as refugees and internally displaced persons (IDPs)
- exploitation of women
- ethnic cleansing.

Geopolitical intervention is controversial even if authorised by the UN. It is argued that the use of military force is justified against a state which is perpetrating human rights violations against its citizens, or in situations of emergency where the aim is to prevent the escalation of humanitarian crises. An example is the 2011 military intervention in Libya. Intervention also takes other formats including economic sanctions and the humanitarian work of NGOs. As shown later, there are many consequences of intervention. Its effectiveness depends on the cooperation and coordination of all organisations involved.

The role of flows of people, money, ideas and technology in geopolitical intervention

All types of intervention involve the exchange of ideas and information. This exchange is an essential part of planning UN missions, including discussion of appropriate strategies by the UN based on data collected. For example the UNHCR employs rapporteurs, special representative and working groups to evaluate human rights issues and develop suitable responses in specific problem areas. It is the policy of many NGOs, for example Amnesty International, to collect information and publicise awareness of human rights abuses in different parts of the world.

The flow of technology into conflict zones includes widespread use of ICT, including use of internet and mobile phones to improve communications. Transfer of information via these modern social media and online tools is important in coordinating activity. Remote sensing through satellite imagery and unmanned aircraft is increasingly important for surveillance of areas that are inaccessible or dangerous.

Biomechanical technology, using the technique of iris scanning, is used in UN refugee camps to register refugees fleeing human rights abuse in conflict zones.

Flows of aid workers and uniformed personnel into conflict zones, as part of UN peacekeeping, amount to many thousands of people per mission. They are involved in a range of tasks from military protection to relief aid in local communities. The flow of money in support of intervention is provided by member states of the UN, charitable donations, foreign aid from governments, and the direct financial support of the IMF.

Knowledge check 36

What role do flows of technology play in geopolitical intervention?

Global governance of human rights involves cooperation and partnership

How human rights are promoted and protected

Strategies to promote and protect human rights are established by organisations which function at different scales, including:

- supra-national such as the UN
- regional such as ASEAN
- national governments
- NGOs, often international in structure but also operational at local scale

The United Nations

The UN is an inter-governmental organisation of 193 member states, each of which accepts the obligations of the UN Charter. Human rights are at its core. The Office of the UN High Commissioner for Human Rights (OHCHR) is the lead agency, based in Geneva, and the UN Security Council, based in New York, deals with serious human rights violations in conflict areas. The UN is a complex organisation. There are many UN bodies relating to specific human rights. Its overall structure can be seen at:

www.un.org/en/aboutun/structure/pdfs/UN_System_Chart_30June2015.pdf.

The UN establishes peacekeeping missions, it may sanction military intervention and it coordinates the input of other organisations in areas of conflict. Further specific examples of its role should be evident in each of your chosen case studies on human rights.

Figure 15 Palais des Nations, Geneva — UN European Headquarters and home to the Office of the UN High Commissioner for Human Rights. This represents the main example of supra-national organisations involved in global governance of human rights issues

> **Exam tip**
>
> Effective intervention requires the cooperation and coordination of all international bodies concerned. Remember that the strategies for intervention do not just include the military option but also humanitarian aid, the mediation and relief work of NGOs, and the application of treaties and international laws.

> **Exam tip**
>
> It is important to understand that intervention by the international community in human rights issues involves many different strategies. These include the input of international organisations, especially the UN, and regional organisations such as NATO, as well as the wide-ranging work of many civil society organisations.

Non-government organisations

Non-government organisations (NGOs) are part of civil society. Their work within conflict zones with local communities is of utmost importance in resolving human rights issues. Their role includes:

- monitoring and providing early warning of new violence
- modification of entrenched and restrictive norms of behaviour through education
- training in particular skills such as agricultural practices and water conservation
- provision of medicines, medical assistance and health education
- supporting women and children's rights.

Treaties and laws

Treaties are formally written agreements between countries, often drawn up with the help of the UN, which are binding in international and national law. Examples of specific national laws should be evident in each of your chosen case studies on human rights. Effective protection of human rights requires coordination of strategies and is based not only on the framework of treaties and international law but also on their practical application and reinforcement through education provided by the personnel of NGOs, the UN and other regional organisations such as ASEAN and NATO.

Knowledge check 37

What is meant by CSOs?

Exam tip

You should be able to evaluate the contribution and roles of the UN and other international organisations in the global governance of human rights.

Knowledge check 38

What is the role of NGOs in the global governance of human rights issues?

Case study of an area of conflict

This part of the specification requires a case study of one area of conflict to illustrate strategies for global governance of human rights and their impact on local communities. Possible choice of country could include Afghanistan or South Sudan.

For your chosen case study you should be able to illustrate the following.

- The contributions and interactions of different organisations at a range of scales from global to local. This should include the work of the UN in the area, one of the national governments involved, and an NGO. This is likely to include the specific strategies of each of these organisations and how they interact, often in partnership, to promote respect for human rights. The strategies might include how these organisations attempt to protect local communities, strengthen rule of law, comply with treaties, promote democratic processes, and protect rights of women and children and those of IDPs and returning refugees.

- The consequences of global governance of human rights for local communities. This could include communities in rural areas and urban neighbourhoods. Consequences could include benefits in the short term, and the longer-term sustainability of development, such as: provision of shelter, improved access to education and health care, training in agricultural practices, securing safe water supplies, and improving hygiene. Depending on your chosen example, you should investigate where local communities may have experienced negative consequences such as the damaging effects of military intervention, including population displacement and possible further disrespect for human rights.

How has intervention in human rights contributed to development?

Global governance of human rights has consequences for citizens and places. The emphasis placed on dealing with human rights issues within the **United Nations Development Programme (UNDP)** suggests there is a strong link between human rights and development. The programme for achieving the eight MDGs has been extended post-2015 to 17 Sustainable Development Goals (SDGs), which are mostly human rights related, see www.undp.org/content/undp/en/home/sdgoverview/post-2015-development-agenda.html.

For example, SDG3, Good Health, SDG4 Quality Education, and SDG5 Gender Equality are three elements of the 2030 Agenda for Sustainable Development.

Global governance of human rights issues affects citizens and places

The intention of global governance of human rights issues is to create benefits not only for individuals and local communities in the short term but also to achieve long-term sustainable socio-economic development. While many effects are beneficial, some attempts at intervention, especially through the use of military force, may lead to unintended negative consequences.

Long-term effects

Long-term effects include positive impacts on development such as the following.

- Improvement in health, increasing life expectancy and reducing IMRs and MMRs.
- Education equality, increasing the enrolment for girls and boys at both primary and secondary level.
- Improved transport systems which provide physical access to services and benefit trade.
- Development of infrastructure networks such as communications.
- Internalisation of more accepted societal norms, especially in strongly patriarchal societies.
- Freedom from abuse of women and children and greater female empowerment.
- Democratic elections, democratic government and political stability.
- Strengthened judicial systems including new national laws and stronger rule of law.
- Employment opportunities and reduction of poverty.
- Development of local agricultural systems including skills training and education.

Short-term effects

Short-term effects include benefits such as:

- medical assistance and provision of medicines through NGOs such as Médicins Sans Frontières
- provision of shelter, sanitation, food and water through the work of NGOs such as ICRC, Oxfam, Save the Children
- military protection preventing further casualties and providing protected areas for civilians to live in and safety for aid workers, which is one of the aims of UN peacekeeping operations

> **Knowledge check 39**
>
> Identify short-term benefits of intervention in human rights issues.

> **Knowledge check 40**
>
> What are the negative impacts of military intervention over human rights issues?

and negative effects of military intervention such as:

- damage to property and infrastructure including hospitals and communications
- population displacement
- further disrespect for human rights
- civilian casualties
- disruption of education as schools and homes are damaged and many children and teachers become refugees
- tensions fuelled over supply of aid, prolonging conflict into the longer term
- military action and ensuing dependence on aid undermining the local agricultural economy.

Exam tip

Remember that you need to know the consequences of global governance of human rights issues. These create: **opportunities** for relief in the short term and for stability, economic growth and development in the longer term, and **challenges** such as the negative impacts of military action.

Case study of one LIDC

This part of the specification requires a case study of an LIDC to illustrate: human rights issues, strategies of global governance and resulting opportunities and challenges. Possible choice of country could include an example from central America and the Caribbean such as Honduras or Haiti, from southeast Asia such as Myanmar or Cambodia, or from central Africa such as the Democratic Republic of the Congo or the Central African Republic.

For your chosen case study you should be able to illustrate the following.

- Human rights issues. These could include effects of issues such as corruption, land-grabbing, discrimination against indigenous populations, women and ethnic groups, human trafficking for child labour, violence against women and poor access to education and health care.

- Global governance strategies. These could include UN involvement such as strengthening government institutions, building a stronger human rights culture, coordinating the work of civil society including NGOs and the involvement of the national government, use of foreign aid and encouraging the Corporate Social Responsibility of MNCs.
- Opportunities. These should include opportunities to improve stability, such as political stability and mechanisms to prevent crime, to promote economic growth including trade agreements and industrial diversification, and encourage development, for example meeting MDG and SDG targets such as improving health and nutrition.
- Challenges. These should include the remaining challenges of inequality, for example in housing, sanitation, access to education and health care, and gender, and injustice such as the effects of discrimination, violence and poverty.

Summary

- International human rights norms are set out as statements in the UDHR. These norms are derived from the long-established customs and ways of living, common to all societies across the world. Even though there are growing numbers of human rights norms, laws and treaties there is still significant global variation in their application and acceptance.
- Understanding terminology such as norms, intervention, global governance and geopolitics is fundamental in appreciating that human rights issues are complex.
- There are global inequalities in the spatial patterns of human rights violations. These variations are influenced by economic, social, political and environmental factors.
- Patterns of gender inequality can be demonstrated by global and national variations in educational opportunities, access to reproductive health services and employment opportunities. These issues can be exemplified by a case study of a country such as India or Afghanistan to illustrate and explain the complexity and contested nature of its specific gender inequality issues, their impact on society, and strategies used to address them.
- There are strong relationships between gender inequality, female empowerment and development.

- The violation of human rights can be both cause and consequence of conflict.
- Effective global governance of human rights issues requires the cooperation and coordination of international organisations including the UN, regional bodies such as NATO and ASEAN, and a number of NGOs and private organisations such as GAVI. The strategies can be exemplified by a case study of an area of conflict such as Afghanistan or South Sudan.
- Strategies for the global governance of human rights involve flows of people, money, ideas and technology.
- Global governance of human rights issues has consequences for people and places. These include opportunities created for stability, economic growth and development, and the remaining challenges of inequality and injustice. In the long term there are intended positive impacts of global governance on the development process. In the short term there are both benefits and unintended negative impacts, especially where military intervention is deployed. These consequences can be exemplified by a case study of an LIDC such as Honduras, Myanmar or the Democratic Republic of the Congo.

Power and borders

What are sovereignty and territorial integrity?

The world political map

The world political map shows boundaries and territories of sovereign nation-states. These political units are the dominant entity of the **global political system**.

The formation of new countries and changes in state boundaries demonstrate that the world political map is dynamic. Examples of change include the secession of:

- South Sudan from Sudan in 2011, creating the world's newest country
- 15 new countries in eastern Europe and central Asia from the former USSR in 1991.

These political changes are not just a matter of territory. They affect **sovereignty** over populations and physical resources, and they have a significant impact on

Secession is the transfer of part of a state's area and population to another state.

the economy and social geography of each area, including global patterns of trade and migration.

Some **international borders** are disputed, for example the claims of India, Pakistan and China in the Jammu and Kashmir area (Figure 16). The reasons for these disputes include cultural, religious and ethnic differences, but access to resources, such as water, is a very important factor.

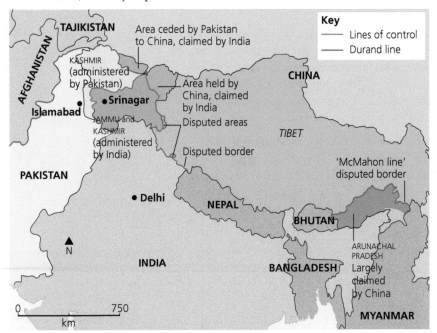

Figure 16 Disputed borders in south Asia

In addition to the **disintegration** caused by secession, change to the world political map has occurred through **integration** by the political and economic grouping of countries. Examples include the formation of regional trading blocs such as the EU and ASEAN, and global organisations such as the UN and G20.

State, nation, sovereignty and territorial integrity

State

A **state** is the area of land of an independent country, with well-defined boundaries, within which there is a politically organised body of people under a single government.

States have the following characteristics:
- defined territory which is internationally recognised
- sovereignty, asserted by the government throughout its bounded territory
- government recognised by other states, often achieved through membership of the UN
- capacity to engage in formal international relations
- **independence**, so that the state is self-governing
- a permanent population which has the right to self-determination.

Exam tip

Correct use of terminology will help your answers to be more concise and authoritative. You should be able to demonstrate clear understanding of terms such as 'norms', 'intervention', 'geopolitics' and 'global governance'.

Knowledge check 41

What is the difference between a state and a nation?

Self-determination is the right of a group with a distinctive territorial identity to freely determine its political status and freely pursue its economic, social and cultural development.

Globally there is inequality in the political power and influence of states. This is closely related to economic and military strength.

The set of institutions and organisations through which state power is achieved is known as the **state apparatus**. These include the internal political organisation, the strength of the legal mechanisms, the organisation of police and armed forces, the provision of health, education and welfare, and the ability to regulate fiscal and monetary arrangements.

State power depends on a range of economic, social, political and physical factors which determine the effectiveness of the state apparatus. For any individual state, examples might include the ability to exploit natural resources, the strength of its international trade and ability to gain access to global supply chains, its human resources including levels of education and demographic structure, industrial development, international relations and government policies.

These factors have a bearing on a state's fragility/resilience. **State fragility** is measured by the Fund for Peace State Fragility Index (Table 9). Examples of indices used include: refugees per capita, fatalities from conflict and number of political prisoners. In 2015, South Sudan, Somalia and the Central African Republic were the most fragile states and Finland, Switzerland and Germany, the most resilient.

Table 9 State Fragility Index and HDI, selected countries, 2015

FFP Classification	Global rank	State	State Fragility Index	HDI
Very high alert	1	South Sudan	114.5	0.467
	2	Somalia	114	n/a
	3	Central African Republic	111.9	0.350
	4	Sudan	110.8	0.473
High alert	5	Congo D.R.	109.7	0.433
	9	Syria	107.9	0.658
	12	Iraq	104.5	0.654
Alert	15	Ivory Coast	100	0.452
Stable	141	Argentina	47.6	0.808
	147	Italy	43.2	0.872
	151	Spain	40.9	0.869
More stable	158	USA	35.3	0.914
	161	UK	33.4	0.892
Sustainable	165	Germany	28.1	0.911
	173	Switzerland	22.3	0.917
Very sustainable	178	Finland	17.8	0.879

[Source: Fund for Peace]

Nation

A **nation** is different to a state. It is a large group of people with strong bonds of identity, which includes their shared descent, history, traditions, culture and language. This may be confined to one country or its people may live in an area across adjoining countries, such as the Kurds, Tuareg or Basques.

Knowledge check 42

Describe the relationship between state fragility and level of development.

Exam tip

In the analysis of statistical maps and diagrams you should be able to describe and explain the patterns and trends shown and evaluate the effectiveness of the technique used to represent the data.

Nations do not have sovereignty. The people, for example the Kurds, may be united in culture but are without a state or sovereign power in an area.

Where a nation has an independent state of its own it is referred to as a **nation-state**. A nation-state is a state that has sovereignty over a single nation. Japan is an example, where the boundaries of the state coincide with the geographical area or territory inhabited by the nation. Many modern states refer to themselves as nation-states even though they may have citizens of more than one nationality. This is partly the result of governments trying to build national identity.

Sovereignty

Sovereignty is the absolute authority exercised by governments of independent states over the land and people in their territory.

The global system of states is based on **territorial sovereignty** which is fundamental in understanding modern political geography. Under this system a state has exclusive authority over its bounded territory and no other state can intervene (internal sovereignty). A state cannot proclaim sovereignty, it has to be recognised by other states. This ensures its territorial integrity and ability to enter international relations (external sovereignty).

These are important principles since, in the globalised world of the twenty-first century, the concepts of sovereignty and territorial integrity are increasingly being challenged.

Territorial integrity

Territorial integrity is the principle that the defined territory of a state, over which it has exclusive and legitimate control, is inviolable.

Territorial integrity and sovereignty are interrelated. Boundaries of states are established by international law. States exercise their authority over this bounded territory.

Maintaining territorial integrity is therefore important in achieving and maintaining international peace, security and stability. According to the Charter of the United Nations, in Article 2.4, member states must not use force or threaten the territorial integrity or political independence of any other state.

Norms, intervention and geopolitics

Norms

Norms are derived from moral principles, customs and ways of living which have developed over time throughout the world and are universally accepted standards of behaviour.

The principles set out in the **Charter of the United Nations** are based on these long-established and universal norms. They are embedded in international law to be upheld by state governments and their citizens.

For example, the entire organisation of the UN is based on the principle of **sovereign equality** of all its member states which is set out in Article 2.1 of the UN Charter. The norms relating to sovereign equality and territorial integrity make it clear that states have the right to self-determination. At the same time, it is a norm that states also have the obligation to protect their citizens, to promote and develop friendly relations, to respect, protect and fulfil human rights, and to allow citizens to be

> **Knowledge check 43**
>
> Distinguish between the terms 'sovereignty' and 'territorial integrity'.

involved in government and have the freedom and opportunity to contribute to society. States which cannot meet the obligations set by these norms are fragile, often because their state apparatus is ineffective.

Norms are reinforced by **international treaties** which are signed by member states. The UN's international treaty collection is an online database which provides information on the status of many multilateral treaties and conventions which have been deposited with the Secretary General of the UN. The treaties or conventions are available in the UN depository for signature and ratification by member states.

There is an increasing number of norms, laws and treaties. These are established globally by the UN and at regional scale in the charters of organisations such as the EU and ASEAN.

Intervention

Intervention is the action of international organisations to resolve conflicts, humanitarian crises and challenges to sovereignty and territorial integrity. Intervention can take the form of:

- economic sanctions, such as a trade embargo
- military action authorised by the UN
- peacekeeping missions by the UN and other regional organisations such as NATO
- humanitarian assistance by the many thousands of **civil society organisations (CSOs)** including NGOs.

Intervention is controversial because the principle of sovereignty promoted by the UN is undermined by the act of intervention, but it may be deemed necessary where:

- a state government fails to protect its citizens from violation of human rights
- there is a direct act of aggression by another state, perhaps over territorial claims
- civil war is the result of poor or corrupt governance
- there is conflict between ethnic groups
- religious fundamentalism or terrorism have serious impacts
- TNCs have negative economic, social or environmental impacts on host countries.

Geopolitics

Geopolitics is the global balance of political power and international relations. The balance of geopolitical power is very uneven throughout the world. The inequalities are explained by the following.

- The power of **advanced countries (ACs)**, such as the USA superpower, which have wealth, high levels of development and politically strong governments. They are in a strong position to drive global systems such as international trade and international migration.
- Peripheral **low income developing countries (LIDCs)** such as Sierra Leone or Lao PDR which are much less powerful, have limited access to global markets and limited control over international migration.
- The rapidly growing economies of **emerging and developing countries (EDCs)** such as India and Brazil are increasingly powerful economically and politically.

Knowledge check 44

What do you understand by the term 'norm' in relation to sovereignty and territorial integrity?

Exam tip

It is important to understand the terms 'state', 'nation', 'sovereignty' and 'territorial integrity' in explaining the global political system and the pattern of current political boundaries.

- The formation of supra-national political and economic organisations such as the UN, EU and OPEC which are able to exert strong geopolitical influence.
- The increasing impact of **transnational corporations (TNCs)** in many countries as globalisation spreads and intensifies.

The **geopolitics of intervention** is very significant. When the **international community** is called upon to intervene there must be consideration of the need for intervention, the type of intervention, the political composition of the groups of countries involved, the characteristics of the country, government and people likely to be affected, and the potential socio-economic, environmental and political impacts.

Global governance of sovereignty and territorial integrity is a term encompassing all types of intervention: economic, military, humanitarian aid, education and reinforcement of norms, laws and treaties. The effectiveness of global governance depends on the interaction, cooperation and coordination of all organisations involved at every scale.

Challenges to sovereignty and territorial integrity

What are the contemporary challenges to sovereign state authority?

Factors influencing erosion of sovereignty and loss of territorial integrity

Current political boundaries

The long-established global political system of nation-states is based on the principles of sovereignty, territorial integrity and sovereign equality of all states as established in the Westphalian model. These principles are reinforced in the Charter of the United Nations. But in the twenty-first century control of territory and a number of international boundaries have been challenged by disruptive and destabilising forces. Examples of these forces include:

- contested territory, such as the Russian annexation of Crimea
- separatism, such as claims for secession by Catalans in Spain
- factional or sectarian tensions, such as political and ethnic conflict in the Middle East
- transnational movement of terrorism and religious extremists
- disputed maritime boundaries, where rights over natural resources and exploration are contested
- the legacy of colonialism where arbitrary political boundaries have caused partition of ethnic groups.

Transnational corporations (TNCs)

Transnational corporations are large corporate enterprises which operate in more than one country. Many are based in ACs, but the number of TNCs is growing as foreign direct investment from EDCs increases.

TNCs have become a driving force of global economic integration. Benefits are created for countries which receive this investment. LIDCs in particular have become more integrated into the global economy.

Knowledge check 45

What is meant by the resilience of a state?

Exam tip

Challenges to sovereignty and territorial integrity are explained by a range of economic, social, political and environmental factors. You should be able to demonstrate how these factors are important, often in combination.

The **Peace of Westphalia** (1648) marked the formal recognition of states as sovereign and independent political entities. It established the principle of sovereign equality of states, forming the basis of international law that governs the global political system today.

But TNCs have been criticised for the disadvantages they can bring which in effect challenge government control and state sovereignty in some countries. Many small, peripheral economies have become dependent on the economic and social benefits derived from TNC investment.

In the twenty-first century, many TNCs have expanded regardless of state boundaries. Such is the economic power of these large companies that some LIDCs have in part lost control of territory, workforce and environment and in some instances their own political decision making. This can affect the lives of many people, especially if a TNC decides to disinvest or if it exploits the workforce. Payment of low wages, demands for overtime and long hours, poor working conditions and use of child labour are all examples of human rights abuses levelled at some large corporations. This disrespect of human rights is a further challenge to the sovereignty of the host country.

These challenges are a concern to many international organisations including the UN and OECD (Organisation for Economic Co-operation and Development). They have established guidelines which set out responsibilities of TNCs and state governments involved. A number of TNCs have responded by making policies with the aim of achieving their Corporate Social Responsibility and conforming to the UN Global Compact.

Supra-national institutions

Supra-national institutions such as the EU or UN represent a tier of government above that of the individual state. Member states retain their sovereignty including their independence, equality, territorial integrity, and responsibility for their citizens.

Nevertheless, states are also bound by the requirements of the supra-national body including any treaties they sign. In this respect they are said to 'surrender' some aspects of their sovereignty since they must comply with the international and regional laws of these institutions.

For example, membership of the EU regional trading bloc affords many benefits of integration which protect state interests such as protection of domestic industry by the common trade tariff and gaining access to the large European market. But these same states have to implement EU laws even if they did not vote for them. They may be bound by stipulations relating to important economic activities such as the Common Fisheries Policy or the Common Agricultural Policy even if they are not totally acceptable in every respect in every part of the EU.

Membership of the UN includes 193 sovereign states. Each state retains its independence, sovereignty, territorial integrity and responsibility for its citizens. These are founding principles of the UN. What appears to conflict with the concept of sovereignty is that the UN Security Council has the right to sanction intervention in a particular country without the consent of its government if it has allowed human rights violations or if it has committed them itself. Any intervention in a state by the international community in these circumstances is controversial since it conflicts with and challenges the concept of sovereignty.

The **UN Global Compact** is an initiative which invites companies to align their strategies and operations according to universal principles on human rights, labour, environment and anti-corruption, and to take actions that advance societal goals.

Exam tip

You should understand that the relationship between patterns of state fragility and levels of development are explained by economic, social, political and environmental factors and that these factors are interrelated.

Political dominance of ethnic groups

The geographical distribution of ethnic groups does not always coincide with current political borders. Challenges to sovereignty and territorial integrity can be explained as follows.

- An ethnic group is partitioned by current state borders. These borders may bear no resemblance to the traditional territory of the ethnic group. Both sovereignty and territorial integrity are challenged when the ethnic group, having strong identity, culture and political organisation, demands full independence to create a new state. Examples include the Tuareg who claim the right to self-determination in Azawad (Figure 17), and the Basques who have a strong tradition of independence in their historic provinces of northern Spain and southwest France.

- A sovereign state includes more than one ethnic group. Sovereignty is challenged when internal conflict between ethnic groups is so violent and widespread that it results in a government being unable to protect all its citizens. An example is the conflict between the Dinka and Nuer in South Sudan.

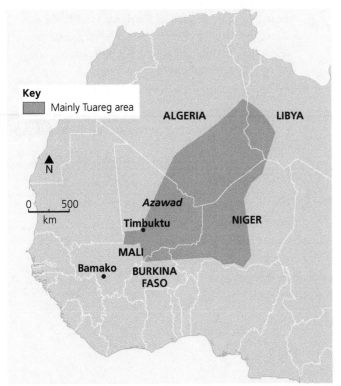

Figure 17 Tribal lands of the Tuareg in Sahelian and Saharan Africa

Exam tip

You should be prepared to evaluate the relative importance of the different factors which cause conflict when sovereignty and territorial integrity are challenged.

Case study of a country where sovereignty has been challenged

This part of the specification requires a case study of any one country where sovereignty has been challenged to illustrate causes, challenges to the government, and impacts on people and places. Possible choice of country could include Ukraine, South Sudan, Mali, or Spain.

For your chosen case study you should investigate and be able to illustrate the following.

■ Causes of the challenge to sovereignty. These could include political factors such as contested political boundaries or political dominance of ethnic groups seeking independence, and economic factors such as the impacts of TNCs or the effects of membership of an economic/political union such as the EU. Factors such as the cultural and political diversity, and the contribution of the country's physical, social and economic geography to this diversity, could be investigated.

■ Challenges to the government. These could include 'threats from within' such as violence or protest in claiming separatism, and ethnic conflict, and 'external threats' to territorial integrity such as invasion or conflict over disputed boundaries or resources with a neighbouring state. In addition, challenges to the government might arise from the fragility of the state apparatus which has led to criminal activity, corruption or an electoral system which is unfair or unreliable.

■ Impacts on people and places. These could include military activity leading to displacement of population, deaths and injuries, damage to housing, poor access to services, medicines and food, the effects of terrorism, disruption to the economy, livelihoods and communications, and loss of power and energy supplies.

The role of global governance in conflict

Global governance provides a framework to regulate the challenge of conflict.

Challenges to sovereignty and territorial integrity can cause conflict

Challenges to sovereignty and territorial integrity which cause conflict include:

■ unjust treatment of citizens, such as limited opportunity to be represented in government
■ competition for access to natural resources, such as water supply
■ suppression or marginalisation of people seeking independence or autonomy
■ where a government fails to protect its citizens from violation of human rights
■ persecution of people for their religious or political beliefs
■ ethnic conflict within a state
■ government failure to ensure supply of basic human needs such as health care, hygiene, education and food.

The global pattern of peace is very uneven. A useful source of data is the **Global Peace Index** produced by the Institute for Economics and Peace, a composite index based on 23 different indices.

One example is the sovereignty of Kashmir which is contested by Pakistan and India who are in conflict over water supply from headwaters of the Indus. This dispute is becoming very important for both countries because:

■ their fast growing populations create increasing demand for water
■ the water resource itself is depleting as global warming causes Himalayan glaciers such as the Siachen Glacier, a major source of the Indus, to retreat.

Knowledge check 47

What do you understand by the term 'global governance'?

The role of institutions, treaties, laws and norms

Threats to the global system of sovereign nation-states may lead to intervention by the international community. This could involve a range of institutions, for example the UN, NATO, the EU and international NGOs. These all have different roles and are involved in establishing and reinforcing treaties, laws and norms which are also important in regulating conflict and in reproducing or maintaining the global system of sovereign nation-states.

The United Nations

The United Nations (UN) was founded in 1945, replacing its forerunner, the League of Nations. It is an international organisation of 193 elected member states. Its overall structure can be seen at: www.un.org/en/aboutun/structure/pdfs/UN_System_Chart_30June2015.pdf.

In regulating conflict, the aims of the UN are to achieve worldwide peace and security, develop good relations between countries, and foster cooperation among nations.

The UN Security Council has primary responsibility for maintaining international peace and security. Figure 18 identifies the ways in which it does so.

Knowledge check 48

What is meant by the international community?

Exam tip

It is important to understand that intervention by the international community in sovereignty and territorial integrity issues involves many different strategies. These include the input of international organisations, especially the UN, and regional organisations such as NATO, as well as the wide-ranging work of many civil society organisations.

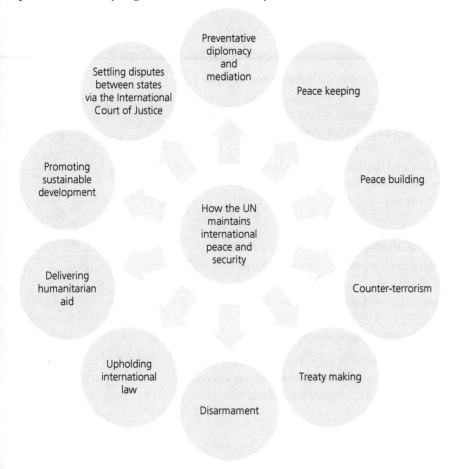

Figure 18 How the UN maintains international peace and security

Some of these practices are becoming more difficult to achieve in the twenty-first century. Therefore the UN increasingly adopts strategies of prevention such as diplomacy, mediation, monitoring and observation to enable its peacekeeping missions to be deployed earlier.

The UN operates all its policies as part of a global partnership. This helps to give any intervention its legitimacy, sustainability and global reach. Globally there are 16 current peacekeeping missions. These involve the input of many UN departments, the support of the host country, and the use of personnel and financial contributions from member states. The UN also coordinates the input of other organisations in areas of conflict. Further specific examples of its role should be evident in each of your chosen case studies on Power and borders.

The North Atlantic Treaty Organisation

The North Atlantic Treaty Organisation (NATO) is an alliance of 28 member states in Europe and North America. It aims to safeguard its members:

- politically, by promoting democratic values and by encouraging consultation and cooperation on defence and security to prevent conflict
- militarily — if diplomatic measures are ineffective it has the military capacity and mandate needed to undertake crisis management operations either alone or in cooperation with other countries and international organisations.

The European Union

The European Union (EU) aims to:

- avoid conflict among its member states and to encourage cooperation by enhancing **economic interdependence** through trade
- provide security, for example through rapid response of OSCE (Organisation for Security and Co-operation in Europe) forces where necessary
- regulate conflict through policies such as its Foreign Affairs and Security Policy, Common Security and Defence Policy and the European Neighbourhood Policy by which the EU aims to cooperate with its close neighbour states in Eastern Europe in terms of security, stability and prosperity.

Non-government organisations

Non-government organisation (NGO) intervention in conflict zones includes:

- monitoring the situation and providing early warning of any new violence
- direct mediation between adversarial parties
- strengthening local institutions, rule of law and democratic processes such as elections
- reinforcing norms, treaties and laws through education at the local scale.

The African Centre for the Constructive Resolution of Disputes (ACCORD) is an example of an NGO specialising in conflict management, analysis and prevention in Africa. NGOs can be particularly effective in assisting individuals, families and local community groups.

Treaties, laws and norms

Treaties, laws and norms are formulated by international institutions such as the UN, EU and ASEAN in order to regulate conflict and maintain peace.

A **treaty** or convention is a written international agreement between two or more states and international organisations. States which sign and ratify a treaty are bound to it by **international law**.

An example directly related to conflict is the *Convention on the prohibition of the use, stockpiling, production and transfer of anti-personnel mines and on their destruction*. The Broken Chair sculpture in Geneva is a monument to the victims and a reminder to all countries to sign the treaty (Figure 19).

Figure 19 The Broken Chair, Place des Nations, Geneva. This illustrates one example of the many treaties which are an important and integral part of global governance of conflict

The UN aims to develop international law based on the multilateral treaties it has adopted. In terms of conflict there are laws on human rights, disarmament, refugees, treatment of prisoners, use of force and conduct of war. You should investigate specific examples in your chosen Power and borders case studies.

Conflict over the **global commons** is also regulated by international law. These include the high seas, the atmosphere, Antarctica and outer space, all of which lie outside the political reach of any one nation-state. There are legal frameworks to address environmental issues, for example the *UN Convention on the Law of the Sea (UNCOL)*.

Treaties and laws are derived from norms. These long-established practices which are common behaviour in many countries are set out in the UN Charter. In turn they are formalised and reinforced by treaty and legal requirements.

Microsoft has created six new norms to deal with the growing threat of **cyber conflict**. This US multinational technology corporation has assumed responsibility for creating cyber security norms for state governments.

Exam tip

You should be able to evaluate the contribution and roles of the UN and other international organisations in the global governance of sovereignty and territorial integrity issues.

Knowledge check 49

Outline the main ways in which the United Nations is involved in resolving sovereignty and territorial integrity issues.

The role of flows of people, money, ideas and technology in geopolitical intervention

Attempts by the international community to intervene in conflict zones include UN missions, involvement by regional organisations such as NATO and the input of CSOs including the many NGOs.

The strategies of these organisations require flows of people and money into the areas of conflict, and their effectiveness is enhanced by flows of ideas and technology. These include:

- movement of personnel into conflict zones on peacekeeping missions
- transfer of finances, including aid often donated by member states
- exchange of ideas in consultation over strategies and flows of intelligence between organisations involved as they attempt to coordinate effective governance
- increasing use of technology such as satellite imagery, remotely controlled drones, weaponry and the ICT required for international databases and communications including mobile social networking.

Exam tip

Effective intervention requires the cooperation and coordination of all international bodies concerned. Remember that the strategies for intervention do not just include the military option but also humanitarian aid, the mediation and relief work of NGOs, and the application of treaties and international laws.

Global governance of conflict involves cooperation and partnership

Case study of strategies for global governance in an area of conflict

This part of the specification requires a case study of one area of conflict to illustrate strategies, cooperation between different organisations and their impact on local communities. A possible example is South Sudan.

For your chosen case study you should be able to illustrate the following.

- Interventions and interactions of different organisations at a range of scales from global to local. This should include the work of the UN in the area, the national government, and an NGO. This is likely to involve their specific roles and strategies and how they interact to deal with different aspects of the conflict and its

consequences such as peace and security, and issues such as health, food and water, refugees, human trafficking and exploitation.

- The consequences of global governance for local communities. These could include benefits, in the short term such as provision of shelter, food and fresh water, medicines and hygiene, and personal safety, and in the longer term such as training in farming practices, and unintended effects of military intervention, such as increased civilian casualties, population displacement, loss of homes and services, damage to infrastructure, increased violence and escalation of human rights issues.

Exam tip

Case studies are important in the more extended responses. The application of statistical and place-specific detail of your chosen areas or countries is useful in supporting discussion and in reinforcing arguments.

Knowledge check 50

What is the role of NGOs in the global governance of sovereignty and territorial issues?

How effective is global governance of sovereignty and territorial integrity?

Global governance of sovereignty and territorial integrity has consequences for citizens and places.

The consequences of global governance of sovereignty issues

Intervention in response to threats to sovereignty invariably has consequences for state governments, citizens and communities involved. The intended benefits of this global governance include the following.

■ Short-term relief, such as the provision of humanitarian aid and donated funds which may include supply of food, improved access to clean water, provision of medicines and medical treatments, provision of shelter and safe havens for Internally Displaced Persons (IDPs), assistance for vulnerable refugees, and attempts to maintain peace, protect civilians and strengthen rule of law.

■ Long-term effects aimed at sustainable development, such as training in agricultural practices to improve food security, education programmes to reduce risk of further conflict, help post-conflict rehabilitation of communities and support minority groups, changes in political regime which involve building of democratic institutions and supporting democratic and fair elections, providing technical assistance in improving legislative and administrative frameworks, upholding human rights and reinforcing norms, treaties and laws through education and training of police and military, and integrating gender equality into policies and daily practices in the home and workplaces. This involves resilience-building in all aspects of the economic, social and political life of the country, especially in strengthening **state apparatus**.

The consequences of global governance of territorial integrity issues

Intervention in response to threats to territorial integrity has consequences for governments, citizens and local communities involved. The intended benefits of this global governance include the following.

■ Short-term relief, such as negotiating periods of ceasefire, maintaining peace, protection of civilians, assistance and provision of shelter for IDPs, human rights monitoring of the treatment of minority groups and women and children, reducing forced conscription of child soldiers, early warning of renewed conflict and strengthening rule of law.

■ Long-term effects designed to achieve political stability, economic growth and sustainable development, such as improving trade relationships to help reduce impact of economic shock, encouraging economic and political cooperation between countries, improving the business environment by counter corruption and diminishing terrorism finance, supporting transition to democracy and fair elections, improving cyber defence, restoring territory according to international law and re-establishing state authority and building resilience in the state apparatus.

Resilience-building is the means by which a state undergoes transition from a position of fragility to one of greater resilience in which its institutional strength and social cohesion enable the promotion of security and development and ability to respond effectively to shocks.

Knowledge check 51

Identify long-term benefits of intervention in sovereignty and territorial integrity issues.

Sovereignty and territorial integrity are interrelated. For example, when territorial integrity is threatened this is also a challenge to sovereignty. Unintended effects of military intervention in instances where either or both have been challenged may exacerbate already existing inequalities and injustices. For example numbers of civilian casualties increase, population is displaced, with refugees and IDPs, there is physical damage to residential districts, including housing and infrastructure, food security is a problem since farming becomes impossible, education programmes are disrupted and there is often escalation of violence and further human rights abuse.

Case study of one LIDC

This part of the specification requires a case study of an LIDC to illustrate sovereignty or territorial integrity issues, strategies of global governance and resulting opportunities and challenges. A possible example is Mali where sovereignty and territorial integrity are both threatened.

For your chosen case study you should be able to investigate and illustrate the following.

- Sovereignty or territorial integrity issues. These could include partition of ethnic groups by arbitrary colonial boundaries, claims for independence by politically dominant ethnic groups, internal conflict including insurgency or coups d'états, terrorism and ineffective state government, which has possibly marginalised some groups. Physical, social and economic inequalities in the country's geography might be factors that contribute to an understanding of the challenges.
- Global governance strategies. These could include UN involvement such as: peacekeeping missions, strengthening government institutions, building a stronger human rights culture, coordinating the work of regional organisations, civil society including NGOs and involvement of the national government.
- Opportunities. These should include opportunities to improve stability, by strengthening all elements of the state apparatus, and in protecting human rights, to improve economic growth, by perhaps reducing import dependency and gaining access to global supply chains and development, by improving water and food supply and health and education services.
- Challenges. These should include the remaining challenges of inequality, such as cultural, socio-economic, and urban-rural divisions and injustice, such as human trafficking, drug smuggling, corruption, gender inequality, mortality rates and poverty.

Exam tip

Remember that you need to know the consequences of global governance of both sovereignty and territorial integrity issues. These create **opportunities** for relief in the short term and for stability, economic growth and development in the longer term, and **challenges** such as the negative impacts of military action.

Knowledge check 52

Outline the unintended effects of military intervention in areas of conflict where sovereignty and territorial integrity have been challenged.

Summary

- The world political map of sovereign nation-states is dynamic. There have been a number of changes to international boundaries, the creation of new states, and the growth of regional economic and political organisations in the last 20–30 years.
- Meanings of the terms 'state' and 'nation', and the concepts of sovereignty and territorial integrity are important in understanding the global political system.
- There is a close relationship between state fragility / resilience and level of development. Global inequalities in the political power and influence of states is related to their economic and military strength. This in turn depends on the strength of their state apparatus.
- Understanding terminology such as norms, intervention, global governance, and geopolitics is important in appreciating that sovereignty and territorial integrity issues are complex.
- Challenges to sovereignty and territorial integrity include threats to current political boundaries, the impact of TNCs, the influence of supra-national institutions and the political dominance of ethnic groups. This can be exemplified by a case study of a country where sovereignty has been challenged such as Ukraine.
- Challenges to sovereignty and territorial integrity can be a cause of conflict.
- Effective global governance of sovereignty and territorial integrity issues requires the cooperation and coordination of international organisations including the UN, regional bodies such as NATO and the EU, a number of NGOs and other civil society organisations. The strategies can be exemplified by a case study of an area of conflict such as South Sudan.
- Strategies for global governance of sovereignty and territorial integrity involve flows of people, money, ideas and technology.
- Global governance of both sovereignty and territorial integrity has consequences for people and places. In the long term there are intended positive impacts on the development process and in the short term there are benefits for local communities. There are also unintended negative impacts especially where military intervention has been deployed. Sovereignty and territorial integrity are closely related concepts. The consequences of the global governance of either or both can be exemplified by a case study of an LIDC such as Mali.

Questions & Answers

Assessment overview

A-level Papers 1 and 2 are both 1 hour 30 minutes long, each carrying 66 marks. Each paper makes up 22% of the A-level qualification.

Paper 1, Physical systems, consists of two sections: Section A Landscape systems and Section B Earth's life support systems. Paper 2, Human interactions, consists of two sections: Section A Changing spaces; making places and Section B Global connections.

In the Earth's life support systems section, you will have to answer the one question that is set — that is, there are no options to choose from. The question will be a combination of short-answer questions, with sub-parts typically ranging from 2 to 4 marks each, one sub-part worth 10 marks and an extended response question (denoted *) worth 16 marks. The grand total for the Earth's life support systems question is 33 marks. It is recommended that you allocate about half the time of the entire paper to this question, that is 45 minutes.

In the Global connections section, you will have to answer two questions: one on Global systems (**either** Trade in the contemporary world **or** Global migration) and one on Global governance (**either** Human rights **or** Power and borders). One question will be a combination of short-answer questions worth 17 marks, with sub-parts typically ranging from 2 to 8 marks each, and the other will be an extended response question (denoted *) worth 16 marks. The grand total for the Global connections section of the exam is 33 marks, for which it is recommended you allocate about half the time for the entire paper, that is 45 minutes.

Exam tip

It is important to prepare for both short-answer questions and essay questions in your chosen Global connections options as you will not know in advance which format will appear in the exam paper.

Some of the short-answer questions will be data-response based on resources provided in a separate booklet. This could include quantitative data such as statistical maps, diagrams and tables or qualitative information in the form of photographs or brief written extracts. Geographical skills will be assessed within this component and these are identified in the specification.

About this section

The questions below are typical of the style and structure that you can expect to see in the A-level paper. Each question is followed by examiner comments which offer some guidance on question interpretation. Student responses are provided, with detailed examiner comments within each answer, to indicate the strengths and weaknesses of the answer and the number of marks that would be awarded. A final summary comment is also provided giving the total mark and grade standard.

Earth's life support systems

Question 1

(a) Study Table 1, the global reservoirs of water.

Table 1 The global reservoirs of water

Store	Size (km^3 x 10^3)	% of global water
Oceans	137,0000	97
Polar ice and glaciers	29,000	2
Groundwater (aquifers)	9,500	0.7
Lakes	125	0.01
Soils	65	0.005
Atmosphere	13	0.001
Rivers	1.7	0.0001
Biosphere	0.6	0.00004

Suggest how the distribution of water amongst the major stores shown in
Table 1 influences human use of **the** water cycle. [4 marks]

(b) Explain three advantages of using satellite surveys in assessing land-use
changes. [3 marks]

(c) Examine the significance of short-term changes in the carbon cycle. [10 marks]

(d)* To what extent can management strategies moderate human impacts on
water and carbon cycles in EITHER the tropical rainforest OR the Arctic tundra? [16 marks]

Total = 33 marks

ⓔ

(a) This sub-part requires interpretation of the table to suggest appropriate uses
of water and justification for the suggestion. Remember to quote figures from
the table.

(b) This sub-part assesses an aspect of geographical skills, in this case the
merits of satellites in data collection about changes in land use.

(c) The third sub-part requires knowledge and understanding of changes in the
carbon cycle. It is important that these are short term and that you interpret
the significance of these changes. This is likely to be more convincing if you
mention examples to support your points.

(d)* The command phrase at the start of the question directs you to make an
assessment which means that you must offer some evaluation or judgement.
You are instructed to set your assessment in one of two specific locations,
therefore you are expected to have substantial knowledge of the management
strategies. The * means that the quality of your extended response will be
assessed in this question. You should use full sentences, spell and punctuate
correctly and make appropriate use of technical terminology.

Student answer

(a) Most of the water is in the oceans and this is where most evaporation comes from. The water in rivers is a very small % of the water (0.0001) [a] but this is vital to humans as it is fresh water and easily used for uses such as irrigation and in homes [b]. Rivers can be dammed and water stored in the lake behind the dam. There is more water in groundwater (0.7%) [a] and humans use this as well. In some places this is a very important source of water as in the dry season rivers dry up and cannot be used [b]. Water can be pumped up from the ground and used.

(e) **4/4 marks awarded** The response focuses on the table and picks up on the use of rivers and groundwater specifically. Figures are quoted accurately [a] from the table, and for both rivers and groundwater, sensible justifications are given for their use by human activities [b].

(b) Satellites are very useful in studying land-use changes. They allow you to see a very large area [a] and this means that you can see changes. For example when a large city has grown, you can see where it has grown and how quickly it has grown. Some satellites go round the Earth but others stay in the same place. Both sorts can show land-use changes.

(e) **1/3 marks awarded** This is a disappointing response as the student is vague about the technique of satellite technology. There is some credit as regards the scale of observation [a] that satellites can operate at but their ability to zoom in to the small scale is not mentioned. The comment about orbiting and geo-stationary satellites has potential as regards explaining an advantage in the mapping of land-use but is not developed.

(c) The carbon cycle is one of the most important cycles on Earth. It is a closed system at the global scale so any changes can affect the global cycle [b]. There are also smaller-scale cycles of carbon which are open systems [b]. For example a single tree is a carbon cycle and carbon goes into the tree and comes out as CO_2 every day [a].

(e) This is an encouraging opening paragraph as it indicates that the student has immediately focused on short-term change [a] in the carbon cycle. It also suggests that the student has grasped the idea of closed and open systems [b].

One key way in which the carbon cycle changes in the short term is through plants. Photosynthesis means that they absorb CO_2 from the atmosphere and accumulate carbon [a]. As plants grow they store carbon in their leaves and stems. Although they also release CO_2 when they respire, they store more carbon than they release [b] and this helps reduce the amount of CO_2 in the atmosphere. This is important as CO_2 is increasing in the atmosphere and this is causing global warming. CO_2 is a greenhouse gas and is trapping more and more of the Sun's energy.

e In this paragraph the student offers thorough knowledge and understanding of the role of plants in the carbon cycle through the process of photosynthesis **a**. With a little more detail about the time scale this cycling operates at, such as the diurnal rhythm of absorbing CO_2 during the day and respiring it at night, the point would be more convincing. The application of this knowledge and understanding to the significance as regards storing carbon is good **b**.

> In the oceans phytoplankton are the same as plants on land as they absorb CO_2 from the atmosphere and store it. This carbon either passes through the food chains such as fish eating phytoplankton and then being eaten by predator fish, such as tuna **a**, or it sinks to the bottom of the oceans when the plankton dies. Both these ways mean that carbon is taken from the atmosphere and stored. The accumulation of dead plankton on the sea bed **b** can then lead to longer-term storage of carbon.

e The student reinforces their response with the comments about phytoplankton, displaying effective knowledge and understanding of the significance of short-term changes in the carbon cycle **a** but also linking this with longer-term change **b**.

> Other short-term changes can be seen when deforestation takes place. Trees contain large amounts of carbon, for example in the rainforest, 60% of all the carbon. When trees are removed this carbon sometimes continues to be stored, for example in timber products such as buildings and furniture **a**. However, if the trees are burned and replaced by either livestock or crops, then the carbon store is severely reduced **a**. Short-term changes can be significant if they end up releasing more carbon than they store and add to the issue of global warming.

e **8/10 marks awarded. Level 3** This final paragraph confirms the student's focus on the question with appropriate comments about deforestation linked directly to short-term changes in the carbon cycle **a**.

The final sentence offers a sensible conclusion to this extended prose response. Overall this answer demonstrates comprehensive knowledge and understanding of short-term carbon cycling although a little more detail would give the discussion greater authority. The student applies their knowledge and understanding thoroughly and links well some of the processes operating in the carbon cycle to the significant issue of global warming.

(d) The water cycle in the Arctic tundra is an interesting one as so much of the water is frozen **a**. Permafrost covers the region and is divided into three types: continuous, discontinuous and sporadic. Which type depends on latitude as the climate is colder the further north one goes. There is a brief summer in the tundra when the temperatures go above 0°C and this is when there is some melting of the permafrost **a**. Liquid water can then flow in the active layer at the surface and water flows into rivers which have a higher discharge.

ⓔ An authoritative paragraph opens this response with the student clearly demonstrating comprehensive knowledge and understanding about some of the factors operating in the Arctic water cycle **a**. This gives the discussion an effective introduction, stating which of the options in the question is being discussed.

> Although humans have lived in the Arctic tundra for thousands of years they have had little impact on the area as the Inuit lived sustainable lives **a**. They basically hunted for their food and in some parts kept reindeer. Recently, human activities have grown in the tundra when oil and gas were discovered at Prudoe Bay in the 1960s. This has meant vast amounts of building such as roads, pipelines and oil rigs as well as houses for the oil workers. This has led to more heat being released into the environment and this has led to the permafrost melting, for example around buildings and pipelines **b**. When the permafrost melts the soil moves and buildings and pipelines slip down slopes. This damages them and they can break and pollute the tundra **b**. It also means that vast amounts of frozen water now flows into rivers and causes floods across the north slope of Alaska.

ⓔ The appreciation that human impacts have been minimal until relatively recently is helpful **a** and indicates the depth of the student's knowledge and understanding and a degree of evaluation. The information about the impacts caused by the oil and gas industries **b** is relevant although the discussion becomes less focused in the comments about slope movements when the permafrost melts. Nevertheless, the student has set the scene well for discussing management strategies.

> There are several strategies which are used to manage impacts on the tundra. One is to put lots of insulation around pipelines **a** in the permafrost and under buildings. This prevents heat escaping and melting the permafrost and the water then is kept frozen. This is an effective strategy although there is another one used with pipelines. These are supported on stilts above the ground **a** which means that any warmth escaping from them goes into the atmosphere as the oil is heated to allow it to flow. In some places such as the Trans-Alaska Pipeline, the supports are refrigerated which means that the permafrost is less likely to melt. It is also important not to melt the tundra as an increase in liquid water means that flooding is more likely. Some parts of the tundra have lots of lakes and ponds caused when gravel was mined for construction. Today permission for this type of mining is not as easy to get so that the permafrost is protected **a**. In the oil and gas industry fewer exploration wells are used as they cause disruption to water movements. Usually underground explosions are used to set off seismic waves which can then be used to tell if the rocks contain oil or gas. They can use computer technology to work out where the oil and gas are. It is also possible to drive a drill sideways so that you don't need to set up a new drilling rig in a different place **a**.

This paragraph is where the student focuses in on the management strategies operating in the Arctic tundra. The student displays comprehensive knowledge and understanding of a number of relevant techniques for avoiding melting the permafrost 🔲. The analysis of these techniques is heading in the right direction but the answer tends towards the descriptive rather than giving a convincing evaluation of the strategies.

> The tundra is a very fragile ecosystem that is very easy to destroy. It takes a long time to recover and so management strategies have to be very good to prevent damage to the water cycle. In the past there was little done to stop damage but now people are more aware of their impacts and more is done to stop environmental damage.

12/16 marks awarded The response ends with appropriate comments about the fragile nature of the Arctic tundra environment. The conclusion also refers to the change in attitudes towards management of the location and assesses the contrast in approach to use of the region.

There is much that is encouraging about this discussion. The student has offered some useful factual material about both the nature of the physical environment of the Arctic tundra, the human impacts and some of the strategies employed to manage these impacts. For the comprehensive knowledge and understanding, Level 3 7/8 marks is given. The key command phrase in the question, 'To what extent...' is addressed but not as comprehensively as it might be to reach Level 3 in AO2. Perhaps some of the time the student spent on the second paragraph could have been given to further analysis of how effective some of the strategies have been. In AO2 therefore, top of Level 2, 5/8 marks.

Total score: 25/33 = on the A/B boundary.

▇ Global connections

Questions 1 and 2 Trade in the contemporary world

1 **Short answer question**
   (a) Study Table 2 an page 94, merchandise exports by value for selected countries, 2014.
      (i) Suggest two reasons for the variation in value of merchandise exports shown in Table 2. [2 marks]
      (ii) Explain one challenge for advanced countries (ACs) shown in Table 2 as result of their international trade. [3 marks]
      (iii) Evaluate one technique which could be used to represent the spatial patterns of merchandise exports by value shown in Table 2. [4 marks]
   (b) With reference to a case study of an emerging and developing country (EDC) explain how international trade contributes to economic development. [8 marks]

Table 2 Merchandise exports by value for selected countries, 2014

Country	Merchandise exports, value in US$ million
China	2,342,306
USA	1,620,532
Germany	1,507,594
UK	505,841
India	321,596
Brazil	225,101
Sweden	164,374
Bulgaria	29,299
Costa Rica	11,252
Zambia	9,696
Lao PDR	2,650
Niger	1,500
Lesotho	925
Afghanistan	571
Central African Rep.	90

[Source: WTO: Trade profiles]

2* **Essay question**

Increasing connectivity in global supply chains depends on modern technology. Discuss.

[16 marks]

1 **(a) (i)** **Two** reasons for the global variation in value of merchandise exports are required. Each reason should be stated clearly with a link to data in Table 2 to reinforce understanding of the variation.

(ii) The question requires identification of **one** challenge. Development of the response should include explanation of how the challenge presents a problem in ACs as a direct result of their international trade.

(iii) This skills question requires choice of **one** appropriate technique and evaluation of its suitability for showing spatial variation in the data in Table 2. The response should include clear identification of the technique and an outline of its advantages and/or disadvantages in representing the data.

(b) This question requires explanation of the beneficial effects of international trade on the economic development of an EDC. All marks are for knowledge and understanding. These can be achieved by making clear links between international trade and economic development in the context of one specific EDC.

2* The command word 'discuss' suggests that this essay requires a discursive response. Discussion should concentrate mainly on the importance of technology and how it helps to increase connectivity in global supply chains.

Understanding would be enhanced by exemplar material where appropriate. While the focus is on technology, the answer should also evaluate the relative importance of other factors on improving global supply chain connectivity. These could include political, economic, social and environmental factors.

Student answers

1 **(a)** **(i)** Differing levels of investment in international transport systems such as port facilities could explain variation in export values for example between China, over 2.3 trillion US$, and Costa Rica, 11.2 billion US$. [a]

e **1/2 marks awarded** The reason given is clearly stated and there is explicit reference to relevant contrasting data in Table 2 which exemplifies the variation [a]. A second reason with some link to the variation would be needed for the second mark.

(ii) Rapid growth of ports in ACs as their international trade expands leads to the challenge of resolving environmental issues such as land-use conflict [a]. For many ACs international trade brings increasing problems of border control, for example in drug smuggling or people trafficking. [a]

e **1/3 marks awarded** In this response, two challenges for ACs as a result of their international trade have been stated [a]. Only one is required. Neither of the stated challenges is developed with explanation.

(iii) Proportional symbols such as circles where the area of the symbol is proportional to the value it represents could be used to show the spatial pattern of merchandise exports. [a] If they were placed on a world map within or near to, the countries shown in Table 2 this would give a very good visual impression of the variations [b] and a clear representation of the geographical distribution. [b] This would be an effective way of representing a wide range of absolute values such as China 2.3 trillion US$ and C.A.R. 90 million US$ on the same map. [b] This is possible because a range of data could be applied to each symbol. There would be little problem of overlap if the circles were correctly placed. But unless specific numbers were added to the map next to each circle, accuracy would be limited to the broad range of values used in the key. [c]

e **4/4 marks awarded** Four relevant evaluative points have been made. Correct choice of representation technique is clearly stated [a]. The technique is evaluated with explicit reference to data in Table 2 with advantages [b] and a potential disadvantage [c]. Construction of the technique is not required per se but aspects of the design may be relevant in support of the evaluative points.

1 **(b)** There is a close relationship between trade and development. The benefits of international trade include economic growth and socio-economic development. In the twenty-first century this has been particularly significant in EDCs such as India.

The Indian government has been actively involved in improving its trade partnerships in order to promote economic development **a**. This has been driven by policies such as removal of high tariffs on imported goods which were in place under its earlier import substitution policy, encouraging both inward and outward FDI, strengthening trade agreements with countries such as the UK with which there is now increasing economic interdependence, and investing in infrastructure and technology.

In 2014, JCB, the British MNC which manufactures construction and agricultural machinery, opened a large factory in Jaipur **b**. This investment can be directly linked to economic development at local scale **c** since it has created jobs, raised incomes and increased spending in the local economy **a**.

At regional scale **c**, agglomeration of industry in the Chennai area is an important economic development which has stimulated the economic multiplier throughout the state of Tamil Nadu and makes a significant contribution to GDP **a**. Chennai has a vehicle-based economy including MNCs such as GM and Ford **b** as well as major Indian companies including Ashok Leyland (commercial vehicles) and Tractors and Farm Equipment Ltd. These have attracted ancillary industries including manufacture of tyres, instruments and electrical equipment **a**. The port of Chennai is a major factor in enabling international trade of vehicles and components.

An important economic development in India is the export of IT services **b**. Global trade in IT services has created billions of dollars in exports annually. Many IT companies have located in cities such as Bangalore creating employment opportunities in a very wide range of skilled and unskilled jobs, and further economic growth **a**.

As a result of international trade in merchandise and services India has experienced significant economic development in the last 15 years. Since 2000 its share of global merchandise exports increased from 0.7% to 1.7% and during the same period HDI has risen from 0.483 to 0.586 **d**.

e **8/8 marks awarded** This response demonstrates thorough knowledge and understanding of the links between international trade and economic development in an EDC, supported by well-developed explanations **a**. There are examples of international trade in FDI, merchandise and services **b** plus reference to their effects on economic development at different scales **c**. This is all set in the context of relevant and substantial knowledge of India, including accurate place-specific detail and statistics **d**. The answer makes good use of terminology. This is a strong answer, placed at the top of Level 3.

e **Total score for the short-answer questions: 14/17 = A grade.**

2* Global supply chains include flows of raw materials, products, services and money in a network of producers, distributors and customers involved in international trade. Supply chains operate between countries of all types of wealth, power and development. They occur within companies such as the large MNCs and between companies. They are important in the global trade system and as the world has become increasingly globalised, supply chains have become more complex and interconnected. A feature of connectivity in supply chains is the role of technology. Technology, though, is not the only influence on supply chains. They are also affected by political, environmental, social and economic factors [d].

Technology in transport is important. Recent growth and development of berthing and handling facilities at ports has revolutionised maritime transport and connectivity in supply chains [a]. The ability to attract large ocean-going vessels has increased not only in ACs but increasingly in EDCs. Examples include ports around the Pacific rim such as Los Angeles, Shanghai and Hong Kong, the ports of southeast Asia including Singapore and South Korean ports, those on the North Sea such as Europoort Rotterdam, Felixstowe and Tilbury. These have become an integral factor in supply chains that have developed within the new trade agreements such as the Trans-Pacific Partnership, the Transatlantic Trade and Investment Partnership, and the EU-South Korea Free Trade Agreement.

The technology to build large ships capable of carrying over 19,000 container units has changed the way in which merchandise is transported within supply chains [a]. ICT at modern ports has involved computerisation of the logistics of moving containers between ports of origin and final destination. It has helped to increase the speed of passage of goods along supply chains and enables their tracking worldwide through use of visibility software [a].

Technology in communications has enabled easier and faster ordering of goods online and making financial transactions [a]. This has helped to bring more producers and customers into global supply chains in the last decade. It has helped speed up transhipments and has minimised the risk of criminal interception. Use of ICT by border and customs agencies has helped administration in supply chains and the governance of corruption [a].

Successful management of global supply chains making use of technology is illustrated by Walmart, the world's biggest retailer and MNC. This American company stocks products in over 70 countries and operates 11,000 stores throughout the world. The geographical diversity of its organisation has necessitated the construction of communication networks to secure easy links and develop strong relationships between suppliers, stockists and customers. This is based on use of technology in the supply chains, including computerisation to achieve the most efficient global transport routes and a tagging system to identify and trace merchandise [c].

Use of technology has been critical in oil supply chains of companies such as Royal Dutch Shell. Technology has enabled development of machinery such as drill bits and platforms for extraction and pipelines for transport of oil in hostile environments such as Alaskan tundra, Middle East deserts or offshore. This includes not only exploration, drilling and transport techniques but also refinery and even planning of temporary accommodation for employees in these source regions. Transhipment and sale of the products have been possible in a growing number of localities because of the role of ICT in efficiency in the supply chains to diverse locations worldwide c.

But technologies are not the only set of factors. Increased supply chain connectivity is achieved by political factors including negotiation of trade agreements between ACs and LIDCs. This integrates companies in LIDCs into supply chains such as the trade links which have been developed between UK and Ghana b. Also opportunities for supply chain development are achieved where countries have joined trading blocs such as the accession of Romania and Bulgaria into the EU, or where political stability has been achieved following periods of conflict and human rights abuses b. These factors help to build the resilience in supply chains to withstand possible economic, political or environmental shocks.

The opportunity to strengthen trade links and to enter supply chains occurs in countries such as EDCs where industrial output has become more diversified or where there is outsourcing of services b. In countries such as India, improving access to education and the raising of skill levels and qualifications of the workforce has enabled productivity to increase. This also assists in gaining access to world markets b. Another contributory factor in the connectivity of supply chains may be the effect of government policy to attract investment in infrastructure b.

Technology is significant in achieving connectivity at each stage of a supply chain. But other socio-economic factors which affect the ability of LIDCs to enter supply chains, and political factors such as negotiation of trade agreements are important too e.

e 13/16 marks awarded This essay is comprehensive in knowledge and understanding of supply chains and the ways in which modern technology can increase their connectivity a. It also identifies factors other than technology which affect supply chain connectivity. This demonstrates convincing analysis of the statement in the question b. It is generalised in places but there is sufficient accurate place-specific detail to reinforce understanding of the main points on technology c. The answer has a well-developed line of reasoning which suggests it was planned. It is clear and logically structured with separate paragraphs for each of the main points. There is a useful introduction which demonstrates understanding of global supply chains and the issues involved d. There is a brief conclusion which makes some attempt to refer to the question e. This essay has scored 7/8 for knowledge and understanding and 6/8 for application of knowledge and understanding.

e Total score for the essay question: 13/16 = A grade.

Questions 3 and 4 Global migration

3 Short answer question

(a) Study Figure 1, the relationship between migrant remittances and HDI for selected countries, 2013.

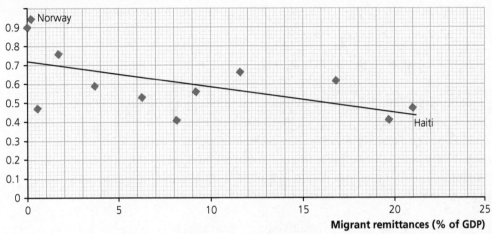

Figure 1 The relationship between migrant remittances and HDI for selected countries, 2013

(i) Suggest two political factors which might affect amounts of migrant remittances in Figure 1. [2 marks]

(ii) Explain one opportunity that migrant remittances can create for LIDCs in Figure 1. [3 marks]

(iii) Evaluate one alternative presentation technique which could be used to represent the data in Figure 1. [4 marks]

(b) With reference to a case study of an advanced country (AC) examine the socio-economic consequences of immigration. [8 marks]

4* Essay question

Economic factors are the most important causes of international migration. Discuss. [16 marks]

3 **(a) (i)** This question requires identification of **two** political factors which might affect level of migrant remittances shown in Figure 1. **One** mark is available for each factor with a clear link to the data in Figure 1.

(ii) The question requires clear identification of one opportunity that migrant remittances can create for LIDCs. For full marks, explanation should be developed so that there is understanding of the link between migrant remittances and the chosen opportunity, with reference to Figure 1.

(iii) This skills question requires the choice of an appropriate alternative technique to represent the two sets of data (HDI and migrant remittances as a percentage of GDP) shown in Figure 1. The chosen technique should be evaluated to show its suitability for representing the data, including the relationship between migrant remittances and HDI.

(b) Socio-economic impacts of immigration on an advanced country are required. These should be supported by specific place knowledge from **one** case study. Some balance of both positive and negative impacts might be expected, but this is not essential for full marks.

4* The assertion that economic factors are the most important in causing international migration can be supported by a range of reasons and examples. Economic migrants represent a significant proportion of current international migration. A good response will also consider other factors which lead to international migration which may be political, social or environmental.

Student answers

3 (a) (i) Membership of a trading bloc, where many economic migrants benefit from free movement of workers, may explain the larger remittances of over 5% **a**. Periods of conflict such as civil war may cause short-term reduction in remittances to less than 1% for some LIDCs since living as refugees there is limited access to work and high costs of travel including payments to traffickers **a**.

ⓔ 2/2 marks awarded This response identifies two factors each of which influences amounts of migrant remittances and is reinforced by interpretation of data in Figure 1 **a**.

(ii) An opportunity is that for many people in LIDCs migrant remittances make an important contribution to the development process, especially in local communities **a**. One reason is that families receiving money sent home by economic migrants use the funds to improve their homes. Over time this can mean more space, better sanitation and improved hygiene **b**. This also provides social benefits such as reduction in spread of disease and contributes to increasing life expectancy **b**. This benefits not only the inhabitants but also the government of the LIDC and can lead to an increase in its HDI.

ⓔ 3/3 marks awarded This answer includes one valid factor which explains the link between migrant remittances and development **a**. Explanation of this link has been developed in an appropriate context of social benefits such as housing quality and health **b**.

(iii) An alternative technique is the use of bar charts – that is two bars for each country placed next to each other, one showing HDI and the other showing migrant remittances as a percentage of GDP. This would demonstrate the overall inverse nature of the relationship between the two sets of data. ACs would have lengthier bars for HDI and shorter bars for remittances, and LIDCs would have shorter bars for HDI and longer for remittances a. Bar graphs are appropriate since they would show each of the two sets of data accurately, especially if numbers were added at the top of the bars a. Visually the differences between ACs and LIDCs would be easy to identify by following the trend in height of the bars a.

e **3/4 marks awarded** This answer includes a brief outline description of the type of bar charts in this instance. This is relevant and provides a basis for understanding the evaluation. Three valid evaluative points are clearly stated a, one of which refers to the relationship between HDI and migrant remittances. A further evaluative point, advantage or disadvantage would be required for full marks.

3 (b) Approximately 41 million immigrants live in the USA, 13% of its total population. The largest diaspora populations are Mexican (12 million), Chinese and Indian (2 million each) c.

Low-skilled Mexicans work in services, construction and agriculture a. This is beneficial to the US economy since they take on the low paid jobs in these sectors which many Americans find unattractive b. High paid, more skilled employment is increasingly being taken up by well qualified immigrants from India and China a meeting shortages in business and science sectors b.

In addition many of the migrants are in the young working age groups a. Not only do they contribute to the US tax base for a number of years b but they also have a demographic impact by increasing the birth rate in an otherwise ageing population b.

There are also problems brought by immigration. It is estimated that there are 11 million illegal entrants, especially from Mexico and south American countries c. This has caused difficulties and expense for the US Department of Homeland Security c in securing land and maritime borders in the southern USA a. This includes construction of a border fence and increased numbers of patrol agents b.

In addition there have been difficulties in integration of immigrant groups into US society a. These arise over issues such as language and qualifications required in job applications and the pressure on supply of resources and services where concentrations of immigrants are high b. This includes Puerto Ricans in New York City and Mexicans in Los Angeles c. Also some immigrant populations are unable to achieve fair and equitable political representation b.

e **6/8 marks awarded** Most of the points demonstrate thorough knowledge and understanding of the socio-economic impacts of immigration in the USA **a**. Explanations are well-developed in the context of the chosen case study **b** although there are some generalisations which could apply to any AC. There is some place-specific detail used to support understanding of the impacts **c**. The response is balanced with reference to both positive and negative impacts of immigration in an advanced country. This response is placed at the lower end of Level 3.

e **Total score for the short-answer questions: 14/17 = A grade.**

4* Economic migrants are by far the largest category of international migrants even though numbers of refugees and asylum seekers have been increasing in the last decade. An economic migrant is someone who makes a permanent or semi-permanent move to another country to improve their standard of living or job opportunity. For these migrants the decision to move is based mainly on economic factors although often a combination of other factors is involved **c**.

This decision is usually a product of their economic status, often one of relative poverty, their employment opportunities, the importance of remittances to the family left at home, and their level of education. These economic factors are not the only causes of international migration. In the case of refugees they are by no means the main factors. There are political, social, and environmental factors which in some instances are more important and necessitate the migration **c**.

The Lee model of migration helps to identify economic factors. This includes push factors, which apply in the country of origin, and pull factors which are attractions of the country of destination. For the potential migrant, some of these factors are real and some are perceived. Many young people, both men and increasingly women as international economic migrants, are 'pushed' away from their country of origin, often an LIDC, by family poverty. This includes socio-economic factors such as overcrowding and poor sanitation in their own homes, lack of access to services, low and unreliable incomes, often a subsistence existence, and limited job opportunities, especially if they are low-skilled or have had limited schooling **a**. For example, these are significant factors that contribute to people leaving west African LIDCs such as Senegal, Liberia or Benin, often in migrations which are intercontinental **d**.

If they can afford to travel, economic migrants are attracted to a more developed country by pull factors which include the real, or at least perceived, economic advantage of employment prospects. Even low paid jobs in ACs often provide much higher incomes than the migrant can earn in the country of origin **a**. For example in two of the largest migration corridors, the USA minimum wage is approximately thirteen times that of Mexico, and in Thailand the minimum wage is now about four times that of Myanmar **d**.

Many potential migrants hear from others who have migrated through the diaspora networks that it is relatively easy and secure using digital technology and the help of banks to send remittances back to their relatives. This money can be used to buy basic equipment for the house or perhaps for farming. The money can be used to improve standards of living by improving the home, for example with latrines or it can be saved for future investment perhaps in a local business such as retail. This is a highly significant factor since the spending in shops can stimulate local economies and the multiplier effect leads to further economic development in otherwise poor communities in LIDCs and EDCs [a]. There are examples of this in rural India and Brazil. Pakistan's emigration policy actually encourages young Pakistanis to work abroad partly to benefit from migrant remittances [d].

The immigration policies in some ACs can be a pull factor designed to satisfy the shortages in their labour force [a]. For example, the Canadian government attempts to attract relatively young workers into the health service, or employment in science, maths and technology sectors, especially in IT. Their points system can be used to fast track graduate and highly skilled economic migrants.

For all these migrants, whether in the short or long term or repeated on a cyclical basis, it is economic factors that explain the move. The main purpose is to improve their way of life and that of their families. But for the refugee other factors assume much more significance especially in the short term. By definition these are people making international migrations because of genuine fear of death or persecution. The push factors include the impact of civil war and threats to homes and livelihoods. Or they may be escaping the strictures of corrupt or undemocratic political regimes where human rights are not respected [b].

In addition there are migrants who move for social reasons such as retirement, as in the case of UK migrants to the Spanish south coast [b]. And there are the migrants seriously affected by the impact of hurricanes, earthquakes and volcanic eruptions on their property and in their communities [b].

It is difficult to be definitive in agreeing with the statement in the question that economic factors are the most important causes of international migration. Globally these are very important and economic migrants are probably the most numerous but there are also political, social and environmental factors which for some people are equally important factors since they can be very pressing in the short term. Moreover, decisions to migrate are often the result of a combination of factors and these depend on the individual circumstances of each potential migrant [c].

ⓔ 14/16 marks awarded This essay is comprehensive in knowledge and understanding of economic factors and the ways in which they lead to international migration **a**. It also identifies other types of factors which cause international migration **b**. There is comprehensive application of knowledge and understanding of the issues, demonstrated through clearly developed and convincing analysis of the statement in the question **c**. Exemplification is generalised in places but there is sufficient place-specific detail to reinforce understanding of the influence of some economic factors **d**. The answer has a well-developed line of reasoning with a clear and logical structure which suggests it was planned. There are separate paragraphs for most of the main points. There is a useful introduction which demonstrates understanding of the issues involved. There is a conclusion which makes an attempt to answer the question.

This essay has scored 7/8 for knowledge and understanding and 7/8 for application of knowledge and understanding.

ⓔ Total score for the essay question: 14/16 = A grade.

Questions 5 and 6 Human rights

5 **Short answer question**

(a) Study Figure 2, the maternal mortality rates for selected countries, 2015.

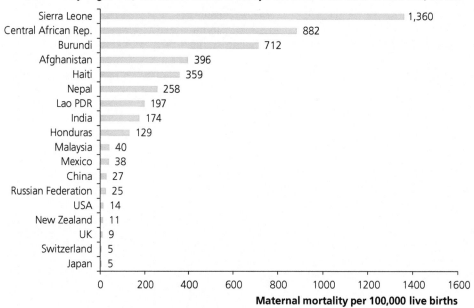

Figure 2 Maternal mortality rates, selected countries, 2015

> **(i)** Suggest two ways in which the status of women in different countries affects variation in maternal mortality rates in Figure 2. [2 marks]
>
> **(ii)** Explain one economic factor which might account for the high maternal mortality rates in Figure 2. [3 marks]

(iii) How effective is the presentation technique used in Figure 2 for showing global variations in maternal mortality rates? [4 marks]

(b) With reference to a case study of an area of conflict, examine the role of NGOs in resolving human rights issues. [8 marks]

6* Essay question

'Strategies to protect the rights of women are effective only in the long term.' To what extent do you agree with this statement? [16 marks]

(e)

5 (a) (i) This question requires application of knowledge and understanding to identify **two** ways in which status of women can be linked to global variation in maternal mortality rates with reference to Figure 2.

(ii) In this question **one** economic factor is required which explains high maternal mortality rates. The factor should be clearly identified, and explanation of the link between the factor and its influence on maternal mortality rates should be developed with explicit reference to the high rates shown in Figure 2.

(iii) This question requires an evaluation of the bar graph technique used in Figure 2 for showing MMR data. For full marks, statements of four advantages and/or disadvantages of the technique are required. There should be explicit reference to the bar graph/data shown in Figure 2.

(b) This question requires an examination of the role of NGOs in resolving human rights issues in a specific area of conflict. All marks are for knowledge and understanding. These can be achieved by making the links between the work of specific NGOs and the human rights issues in the context of one country or area of conflict.

6* This question requires discussion of the strategies used to protect the rights of women, including an evaluation of their effectiveness over time. 'To what extent..?' suggests that while the response should explain strategies which require the longer term to be effective, it should also consider strategies which have more immediate impact. Arguments should be supported by exemplar material.

Student answers

5 (a) (i) Limited female access to education, information and lack of empowerment contribute to high maternal mortality rates, for example in sub-Saharan Africa (CAR, 882/100,000 live births) a. MMRs are much lower where women have greater control over family size or frequency of pregnancies and where attitudes towards women's health are positive, for example in Europe (UK, 9/100,000 live births) a.

(e) **2/2 marks awarded** This response clearly identifies two aspects of the status of women and each is linked to contrasting MMR statistics from Figure 2 which show variation between an LIDC and an AC a.

(ii) Lack of money available for investment in access to medical treatment and training of skilled practitioners can lead to high MMRs in LIDCs [a]. In 2015 this contributed to the very high MMR of 1,360/100,000 live births in Sierra Leone. This economic factor restricts the development of medical facilities, communications and road networks in rural areas which present severe restrictions in the case of emergencies in LIDCs [b]. Inability to invest in training doctors, midwives and paediatricians also adds to the problem by limiting the quality of medical services and the provision of skilled attendants at birth [b].

(e) **3/3 marks awarded** This response includes clear identification of one economic factor [a]. Explanation is developed [b] in the context of high MMRs in LIDCs including a link to the data for Sierra Leone provided in Figure 2.

(iii) The bar graph technique gives effective visual representation of the MMR data. It shows the relative size of the inequalities and allows comparison of countries to be immediately clear [a]. There is accurate representation of the data since the length of the bars in Figure 2 is proportional to the MMR values they represent for each named country. This shows clear global variation between specific LIDCs, EDCs and ACs [a]. Placement of MMR statistics at the end of each bar is an added advantage, for example Central African Republic (882) and Japan (5) . This contributes to the precision of the diagram [a].

(e) **3/4 marks awarded** This answer gives three valid advantages of the bar chart technique [a]. These points are specifically linked to the effectiveness of the diagram for showing global variation in MMR statistics. There is explicit reference to data provided in Figure 2. One further advantage or disadvantage would be needed for full marks.

(b) In areas of conflict such as Afghanistan, rural populations face many different types of human rights issues. Afghan Aid is an NGO specifically working on problems in rural areas. These include discrimination against women, who are denied the same rights as men in access to education, employment and health services. In addition, conflict has led to civilian casualties and deaths so that fear of being caught up in military action, including the impact on children, is an everyday issue of human rights for local communities. Resurgence of the Taliban has led to further human rights violations including the issue of food security as farmland is used for poppy cultivation to fuel the drugs trade. Lack of government investment in roads and communications are further breaches of human rights by limiting access to basic services [c].

Many hundreds of NGOs have registered to work in Afghanistan to help deal with these problems. Other examples include CARE International and ICRC, the International Committee of the Red Cross [c]. Each tends to specialise in a particular area of concern. Afghan Aid has been particularly effective in assisting with rural development projects a in the Ghor area, an inhospitable and increasingly insecure area politically c.

Because of its close contact with local people, Afghan Aid plays a vital role in coordinating the inputs of the UN and the Afghan government in the rural project work a. This includes the allocation of donor funding a. Specifically they have enabled local communities in the Chaghcharan district c to have greater freedom from the restrictions of the Taliban and to have control over their own development a. This is done through education programmes, training in agriculture, securing safe water supplies, improving hygiene and reducing risk of disease b.

Afghan Aid has helped to integrate women into society a and to involve men and women in democratic processes such as election of their own leaders and creating their own local community development plans b. Overall the work of NGOs like Afghan Aid throughout Afghanistan is helping local communities and neighbourhoods to achieve a better quality of life and more sustainable socio-economic development.

e 7/8 marks awarded This response identifies a variety of the roles of NGOs in resolving human rights issues in an area of conflict a. There is indication of how the work of NGOs is achieved b. The input of Afghan Aid is linked to the human rights issues and this demonstrates thorough knowledge and understanding of the role of NGOs in the context of some place-specific detail of the country c. This response is placed in the middle of Level 3.

Total score for the short-answer questions: 15/17 = A grade.

6* There have been advances in the overall protection of human rights this century, but women and girls continue to experience the effects of gender-based discrimination c. This has a personal impact on women and their families and on the entire rate of progress of development in a country, area or local community. The strategies to protect and promote women's rights include governance by global organisations such as the UN, the significant input of many NGOs, private organisations and activists and increasingly the Corporate Social Responsibilities of MNCs. There are many obstacles to remedy in the short term.

Lack of respect for women's rights takes many forms. The issue of personal safety is a growing problem in parts of India where UNICEF has reported increases in domestic violence, dowry killings and rape and violence outside the home a. It is reported that the number of honour killings have increased in Pakistan and India e where young women have refused to agree to arranged marriage. In states such as Bihar women are expected to lead a life of servitude in their own homes and to raise children. In some countries of sub-Saharan Africa the practices of female genital mutilation and 'breast ironing' continue to have a devastating effect on the lives of women a. In almost every continent women and girls are subject to trafficking for modern slavery including sexual exploitation a.

Women's rights are severely affected by gender inequality. Girls have limited access to education. Enrolment into primary education has improved but there are still many challenges with their exclusion beyond this, especially in the rural areas of poor countries a. Women suffer through discrimination in the workplace a. In India many employers do not meet their maternity benefit commitments. In Delhi only 25% of married women return to the workplace after childbirth e. Female participation in the workplace varies throughout the world. It is very low in Iraq and Afghanistan and, although not equal yet, it is much higher in UK and Switzerland for example e. Female access to reproductive health care varies considerably and this is a factor in the high and unacceptable number of female deaths related to pregnancy and childbirth a, for example in the countries of west Africa.

Strategies to protect women's rights include the MDGs within the UN Development Programme b. These have been effective in the longer term. In the 15 years since the MDG for primary education was established in 2000, female enrolment in primary schools has risen to 70% in India e. The SDGs in the UN's sustainable development agenda include Goal 4 with targets for inclusive and equitable education for all by 2030 b. In Goal 5 the aim is to achieve gender equality and female empowerment in the relatively long term of the next 15 years b. But the policies depend on the support of companies and stakeholders for their implementation.

The impact of NGOs can have a more immediate effect in protecting women's rights b. NGOs such as UN Women, Womankind and the International Committee for Research on Women work 'in the field' with individuals, families and local communities to resolve problems, to change attitudes, and to reinforce laws and the stipulations of treaties through education b. For example ICRW working in partnership with UN Women and the Indian government have set up a 'Safe Cities' project in Delhi e. One relatively short-term effect has been to give women greater confidence to report crimes b. In a short period too, UN Habitat working on development projects in urban neighbourhoods of Kabul e has helped to engage Afghan women in local community projects through their fair and equitable election as local council members b.

Rates of FGM have been slow to fall in sub-Saharan Africa. UNICEF and WHO work together with NGOs at grass roots level with local communities. Tostan, a Senegal-based NGO e, finds that reduction of the problem can be faster when using traditional group discussions to educate about the risks of this harmful and dangerous practice b.

Most strategies for protection of women's rights are only effective in the long term because there are many obstacles to short-term remedy d. These include the entrenched social attitudes and patriarchal norms which are difficult to change in rural India for example. The high incidence of discrimination against women reinforces their lack of empowerment. The legal mechanisms in place, such as many of India's laws to prohibit child marriage and dowry or to protect women from domestic violence e are very difficult to enforce b. Levels of poverty, poor communications and lack of knowledge in many poor LIDCs also add to the vulnerability of women and girls b.

e **14/16 marks awarded** This essay demonstrates comprehensive knowledge and understanding of women's rights issues **a**. The effectiveness of strategies to protect women's rights is discussed in long-term and short-term contexts **b**. This demonstrates comprehensive application of knowledge and understanding. The answer has a clear structure with an introduction **c** which shows understanding of the issues involved. The conclusion is a useful summary explaining why some strategies are not effective in the short term **d**. There is accurate place-specific detail applied appropriately to demonstrate a range of countries and policies **e**, although references are confined to LIDCs and EDCs.

This essay has scored 7/8 for knowledge and understanding and 7/8 for application of knowledge and understanding.

e **Total score for the essay question: 14/16 = A grade.**

Questions 7 and 8 Power and borders

7 **Short answer question**

(a) Study Figure 3, north and central America, state fragility index, 2014. The state fragility index measures relative fragility/resilience of states based on social, economic, political and military indices.

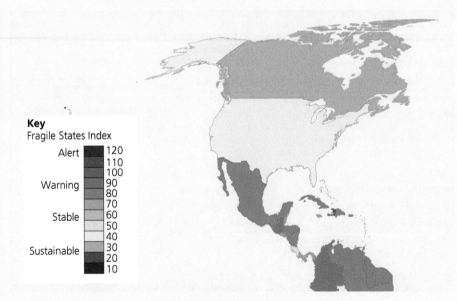

Figure 3 North and central America, state fragility index, 2014

(i) Suggest two ways state fragility shown in Figure 3 can influence flows of people. [2 marks]

(ii) Explain one factor that might account for the spatial variations shown in Figure 3. [3 marks]

(iii) Evaluate the presentation technique used in Figure 3 to show spatial variations in state fragility. [4 marks]

(b) With reference to a case study of an area of conflict, examine the short-term benefits of intervention for local communities. [8 marks]

8* Essay question

'The UN is the most important organisation in the global governance of threats to sovereignty.' To what extent do you agree with this statement? [16 marks]

7 **(a) (i)** This question requires application of knowledge and understanding to show **two** ways in which the state fragility data shown in Figure 3 can influence flows of people.

(ii) The requirement in this question is to demonstrate understanding of one factor which might account for spatial variation in state fragility. The factor should be clearly identified and then explained by further development of the response. Explanation of the spatial variation should be supported by reference to data in Figure 3. Knowledge of state fragility factors for specific countries is not required.

(iii) This question requires an evaluation of the choropleth technique used in Figure 3 to show spatial variation in state fragility. For full marks the answer should include **four** advantages and/or disadvantages. There should be explicit reference to the choropleth map shown in Figure 3.

(b) The response should be set in the context of a named area of conflict, with place-specific detail used to illustrate and reinforce the points made. The requirement of the command 'examine' is to identify and explain the short-term effects of intervention on local communities in the chosen area of conflict.

8* This question requires discussion of the role of the UN in the global governance of threats to sovereignty. The extent to which the UN is the most important organisation in intervention should be evaluated relative to the roles of other organisations in the international community. This will show application of knowledge and understanding. Case study or place-specific detail should be incorporated where appropriate.

Student answers

7 **(a) (i)** The high degree of state fragility in countries such as Haiti (alert category of the fragility index) could lead to significant out-migration to countries of greater political stability, safety and employment opportunities a. Flows of UN personnel on peacekeeping missions and aid workers with NGOs may move from more stable countries such as USA to those with high levels of state fragility in times of crisis in human rights abuse a.

2/2 marks awarded This response identifies two ways in which flows of people are influenced by differing levels of state fragility a. Reference to data in Figure 3 reinforces knowledge and understanding of the link between state fragility and flows of people in each instance.

(ii) Ability to recover from economic shock explains spatial variation in state resilience in north and central America a. Advanced countries such as the USA, in the stable category of the fragility index, are economically more resilient as a result of their economic wealth based on diverse industrial structure and strength in international trade b. Some LIDCs of central America such as Guatemala are in the warning category of state fragility. These are more peripheral economies which are relatively fragile states since they have poorer access to global supply chains and depend on a narrower range of industries b.

e **3/3 marks awarded** One factor is clearly stated – ability to recover from economic shock a. This factor is explained and developed in the context of spatial variation in state fragility with reference to two contrasting countries shown in Figure 3 b.

(iii) The choropleth technique used in Figure 3 is visually effective. The gradation of colours shows clear spatial variation in state fragility between countries and between regions, for example in the contrasts between North America and the Caribbean a. The size of the class interval is an inherent problem for choropleths in that the map does not show precise values for each country, in this case a range of ten index points. Mexico and Venezuela appear to have the same state fragility but they may differ within this range a. Nevertheless the class interval of 10 is relatively small. Therefore, the 12 categories in the key enable small differences in state fragility to be distinguished such as between Guatemala and Honduras a. The block shading of entire countries does have the visual effect of making categories for larger countries such as Canada or USA dominate the map a.

e **4/4 marks awarded** Four points have been made which evaluate the effectiveness of Figure 3 for showing variation in state fragility a. Each point is explicitly linked to evidence on the map or key.

(b) Conflict in South Sudan is the result of clashes between ethnic groups fuelled by the political infighting of the country's leaders. Dinka, Nuer and other ethnic group fighting has caused thousands of deaths since independence in 2011, over 1.5 million IDPs, food insecurity and huge strain on provision of health services.

Intervention has been necessary through the UN and its agencies, WHO, UNICEF and UNHCR in particular. UNMISS, the UN peacekeeping mission has been strengthened recently with additional personnel. IGAD, the Inter-Governmental Authority on Development for eastern Africa, has tried to work in mediation with the government. There are many NGOs working on health, food and water supply and shelter provision for people of local rural communities. The work of all these organisations has been hampered by the obstructive attitude of the South Sudanese government.

People living in small rural communities have experienced some short-term benefits. All-year-round food supply by aid agencies [a] has been made possible by locating food distribution centres where they can be more accessible in the wet season when the roads are impassable [b]. For example, the Oxfam distribution centre at Mingkaman, Lakes State [c]. Relief work by Médecins Sans Frontières [b], although severely restricted by looting and violence has helped in the delivery of babies, treatment of war wounded and treatment for malaria [a] in many areas, for example around Pibor in Jonglei State [c]. Personal safety is an issue in this conflict. UNHCR [b] has set up civilian protection camps [a] for IDPs in Unity and Upper Nile States [c]. WHO [b] has established a cholera treatment centre [a] in response to an outbreak of the disease near Juba [c].

The conflict has severely limited the long-term development process throughout the country. Even the short-term benefits of intervention are very difficult to achieve. Aid organisations are working under difficult and dangerous conditions.

[e] **5/8 marks awarded** Information on the conflict and intervention is useful case study information on South Sudan since it explains how people have been affected. The references to difficulties in provision of aid and peacekeeping are also useful understanding and place-study knowledge. The most direct part of this response is in the third paragraph. This identifies short-term benefits of intervention [a]. For each benefit there is a brief outline of the problem and the way in which it has been achieved [b] plus accurate place-specific detail [c]. More thorough explanation would be needed for a higher level award, but knowledge and understanding is reasonable, and explanation is developed enough to place this response at the upper end of Level 2.

Total mark for the short-answer questions: 14/17 = A grade.

8* Global governance is the term used when the international community is called upon to deal with serious threats to sovereignty and any ensuing humanitarian crisis. The threats to sovereignty can be external, for example when there is an act of aggression possibly over territorial claims, or internal as a result of human rights abuse, claims for independence, ethnic conflict, civil war or insurgency. These situations may require intervention by different types of organisation which tend to have different roles. The UN is an important international organisation nearly always heavily involved. Others include regional organisations such as NATO, plus the many civil society organisations including NGOs and private enterprises [e].

The role of the UN in global governance is varied. In cases of humanitarian crises it may require the UN Security Council to sanction military intervention [a]. This may be thought necessary where severe human rights abuses are perpetrated, such as ethnic cleansing, and where the national government is either unable to protect its own citizens or it is actually allowing the abuse to happen [b]. There was military intervention in Libya in 2011 by a coalition of countries including USA, UK and France following a Security Council resolution. On the same basis there was military intervention in Mali following a coup d'état in 2012 [d].

Intervention by the UN also takes the form of peacekeeping and peacebuilding missions [a]. For example, in Mali, MINUSMA was established to help stabilise the country and to support the political process of re-establishing state authority. The aim was to protect civilians and to promote and protect human rights. The importance of the UN in this example is illustrated by the fact that it was able to involve over 11,000 military, police and other staff drawn from member states [b]. Similarly the UN has intervened in the conflict in South Sudan where sovereignty is threatened by prolonged fighting between ethnic groups as well as the obstructive attitude of the government. Here the UN mission UNMISS was given a new mandate in 2014 to consolidate peace and help with state-building and development. The UN has been able to draw upon 16,000 military, police and other staff from around the world with an approved budget of over US$ 1 billion [d].

The UN also has an important role in provision of humanitarian aid [a]. This is an aspect of intervention conducted by UN bodies such as UNHCR (refugees), UNICEF (children) and UNDP (development) and by associated organisations such as WHO (health). These all play a part in helping civilians in areas of conflict where sovereignty is challenged and the national government is unable to provide protection or basic needs such as food, water, health, shelter and education [b]. This is particularly important in South Sudan where WHO has established a cholera treatment programme, where UNHCR assists over 1.5 million IDPs and refugees and where UNICEF is helping to provide food and water supplies and to assist the many children suffering malnutrition. In this particular case their role is very important in the face of continued fighting, human rights atrocities, conscription of children and an ineffective government [d].

The UN has a vital role in reinforcing the norms in the Charter of the United Nations by treaty-making [a]. There is encouragement for states to sign and ratify treaties which binds them by international law for commitments to peace and security [b]. For example the UN has managed to secure South Sudan's signature of CEDAW and the Convention on the Rights of the Child [d].

While the UN has assumed an important role at all scales, other organisations also have significant inputs in conflict zones and in urgent humanitarian crises resulting from threats to sovereignty. These include regional organisations such as NATO which protects its 28 member states and other areas by its diplomatic role and its potential to deploy force [c]. The EU has rapid response teams within the OSCE organisation, currently involved in monitoring the situation [c] in eastern Ukraine where sovereignty has been threatened by territorial claims of the Russian Federation in Crimea for example [d].

There are many thousands of NGOs such as Médecins Sans Frontières, Oxfam and the International Committee of the Red Cross [d] which are very important in providing humanitarian aid in areas of conflict over sovereignty. They are international organisations but play significant roles 'in the field' working with individuals, families and communities in education, training, securing food and water supplies, medical aid and provision of shelter c. They also are in a good position through co-operation to be effective in the roles of mediation, monitoring and providing early warning of new violence, ensuring local elections are more democratic and, by education, reinforcing norms and human rights c.

It could be argued that even though other organisations such as INGOs play a vital role in global governance, the UN is the most important. It is involved in such a wide range of forms of intervention, it coordinates and encourages cooperation in the work of all other agencies, organisations and governments, and its Security Council can give legitimacy to military action if necessary. The UN also has global reach involving the support of its member states. Moreover in the aftermath of any conflict or crisis it continues to promote the work of development through the UNDP agenda for sustainable development. f

e 14/16 marks awarded The different roles of the UN in intervention over threats to sovereignty have been identified a and their purpose / importance is discussed b. Also the importance of the UN strategies is evaluated in comparison with the role of other organisations c. This demonstrates comprehensive application of knowledge and understanding. Throughout the essay there is appropriate reference to case study / accurate place-specific detail to illustrate and reinforce knowledge of the importance of the strategies d. The answer has a well-developed line of reasoning with separate paragraphs for each point, including an introduction which clarifies the terminology and the issues outlined in the question e. There is a relevant conclusion which gives reasons to reinforce the candidate's overall view that the UN is the most important organisation in global governance of threats to sovereignty f.

This essay has scored 7/8 for knowledge and understanding and 7/8 for application of knowledge and understanding.

Total mark for the essay question: 14/16 = A grade.

Earth's life support systems

1 Biological, e.g. photosynthesis, transporting substances within organisms; environmental, e.g. moderating temperatures by the oceans absorbing, storing, transporting and releasing heat energy; and economic, e.g. power generation, irrigation of crops, use in manufacturing.

2 Water moves rapidly in and out of the atmosphere allowing significant volumes of water to transfer from one location to others such as from oceans to land.

3 Burning fossil fuels releases carbon in the form of CO_2 and so unlocks carbon that otherwise would stay in storage for millions of years. CO_2 is a significant greenhouse gas due to its ability to trap radiation within the atmosphere and contribute to global warming.

4 In the summer, vegetation is in full growth so has the maximum surface area of leaves and stems. Deciduous trees shed their leaves in the winter. The same contrast can be seen in regions where there is a marked dry and wet season such as semi-arid areas.

5 Biomass use leads to no net addition of CO_2 to the atmosphere whereas the burning of fossil fuels releases carbon held for millions of years in very long-term stores.

6 The oceans are where vast amounts of carbon are in long-term storage. Ocean currents carry dissolved carbon to great depths where it stays for centuries. When marine organisms, e.g. phytoplankton and any organisms with shells die they sink to the ocean floor taking with them carbon in their bodies.

7 Extreme events such as flooding can threaten lives in any location. Where rainfall is highly seasonal, e.g. monsoon climate, agriculture can come under serious stress from either flooding or drought. Flooding can also spread water-borne disease and lead to increased risk from insects such as mosquitoes.

8 'Ecosystem services' are benefits (tangible and intangible) that human activities receive from natural environments, e.g. food, water, flood regulation, coastal protection, disease prevention and recreation.

Trade in the contemporary world

9 'Terms of trade' means the value of a country's exports relative to that of its imports. The 'balance of payments' is the difference between a country's inflows and outflows of money, including all transactions with the rest of the world for goods and services, flows of FDI and migrant remittances, over a period of time.

10 FDI is foreign direct investment which is the inward investment by a foreign company (usually a large TNC) in a country.

11 The term 'spatial pattern' refers to a geographical distribution over an area at any scale from global to local. For international trade this could refer to data which has been mapped, for example, on a choropleth such as percentage share in value of global exports by country. Or it could refer to the distribution of container ports in a country.

12 HDI is the Human Development Index which is a composite index of socio-economic development. Its components include economic and social indices. The more developed countries have higher HDI values.

13 'The economic multiplier effect' is the concept that an initial investment in an economic activity in an area has beneficial knock-on effects elsewhere in the area's economy.

14 The development gap is the difference in prosperity and wellbeing between rich and poor countries. This could be measured, for example, by GDP per capita and HDI.

15 Global supply chains are flows of materials, products, information, services and finance in a network of suppliers, manufacturers, distributors and customers around the world. Where value is added to a product by the processes at each stage of a supply chain in different parts of the world it is known as a global value chain.

16 Physical factors influencing spatial patterns of merchandise trade include: the distribution of natural resources such as oil or minerals; climate, soils and water supply which influence supply of specific types of agriculture product; and the natural configuration of a coastline and water depth which may influence location of port facilities.

17 An example of economic interdependence is the mutual dependence of two or more countries in a reciprocal relationship through trade, where a country, such as an AC, exports manufactured goods to another, such as an LIDC, and imports raw materials in return.

18 Foreign exchange generated by international trade can be used in an LIDC to invest in health and education services. FDI can bring benefits where the Corporate Social Responsibilities of MNCs are applied. Bilateral relationships developed with ACs can help to strengthen human rights.

Global migration

19 An economic migrant is a person who moves from another country, region or place, involving a permanent or semi-permanent change of residence, to improve their standard of living or job opportunities. A refugee is a person who has moved outside the country of his/her nationality or usual domicile because of genuine fear of persecution or death.

20 'Net migration gain' is when the number of people moving permanently into an area exceeds the number moving out.

21 Push factors are the negative attributes of a migrant's place of origin which force the migrant to leave. Pull factors are the positive attributes of a place of destination which attract migrants. Often the decision to migrate results from a combination of push and pull factors.

22 A migrant remittance is money transferred from one country to another, sent home by migrants to their family, friends and community. These funds can have significant effects on development, lifting households out of poverty, encouraging spending in the local area and stimulating the multiplier effect. Social remittances introduced by returning migrants include ideas and values, such as attitude towards family size, which also contribute to development.

23 The term 'diaspora' refers to the geographical dispersal or scattering of a population such as an ethnic or national group from their original home. Diaspora communities are part of the networks of peoples which have their origin in one country or area but are now located and identifiable in many countries around the world.

24 In terms of migration, origin refers to the country of their usual residence from which migrants have moved. Destination is the country to which migrants move. This has to be for at least one year for the migrant to be classified by the UN as a long-term migrant.

25 'Internal migration' refers to the migrant flows from one part of a country to another. Rural–urban migrations within a country are significant flows within the global migration system.

26 An asylum seeker is a person who seeks entry to another country by claiming to be a refugee. Their credibility is assessed by authorities in the country to which they have applied for asylum.

27 An example of socio-economic interdependence is the mutual dependence of two or more countries in a reciprocal relationship through migration, where economic migrants provide labour in a country and on return bring newly acquired skills, ideas and values to their country of origin.

28 Challenges for LIDCs arising from international migration include loss of the young, most vibrant element of the workforce and reproductive age groups, loss of the most skilled and better educated of the population in a 'brain drain', and the problem of exploitation of young members of the workforce, especially women and girls by traffickers, including the injustice of forced labour.

Human rights

29 'Human rights norms' are established customary behaviours based on moral principles and ways of living inculcated into the culture of a country or area over a long period of time. Statements set out in the UDHR are drawn from such behaviour across the world and are generally accepted as human rights norms.

30 Intervention by the international community includes the actions of international organisations, or a group of states which have been authorised by the UN Security Council, in a foreign territory to end gross violations of human rights. This includes military force, economic sanctions, and the work of NGOs.

31 Forced labour is when people are coerced to work through use of violence or intimidation, or by more subtle means of detention such as retention of identity papers.

32 An inverse relationship: the higher the maternal mortality rates the lower the level of development.

33 Female empowerment through education of girls and women helps them to move into the labour force of a country or area. It also helps to reduce total fertility rates, infant mortality rates and overall population growth rates. Family health and child nutrition improve and there is often poverty reduction.

34 'Global governance' is the intervention by the global community, in all its formats, attempting to regulate issues, such as human rights, sovereignty and territorial integrity.

35 UN involvement includes its peacekeeping missions in areas of conflict to protect and promote human rights, coordination of the work of other international/regional organisations including NGOs, the establishment and reinforcement of human rights norms, treaties and international laws, and creating sustainable development goals which are mostly human-rights based.

36 Flows of technology into conflict zones include use of social media, online tools, satellite imagery, drone aircraft and modern weaponry. These assist in geopolitical intervention by enabling databases and other information to be collected, to speed communications and intelligence in the coordination of strategies, and for accuracy in surveillance and air strikes in inaccessible and dangerous areas.

37 CSOs are civil society organisations which include NGOs and other private organisations that are independent of governments, working voluntarily, either individually or collectively, in support of citizens and communities throughout the world.

38 NGOs often work in conflict zones to assist local communities. Their role, in cooperation with other organisations, is significant in monitoring violence, education regarding human rights norms and laws, training in agricultural techniques, medical assistance, and reinforcing democratic processes.

39 The short-term benefits of global governance for local communities in conflict areas include provision of medicine and medical treatment, shelter and sanitation, food and water, and military protection.

40 Military intervention can have effects such as causing further civilian casualties, population displacement, damage to property and infrastructure, further disrespect for human rights, disruption of children's education, and dependence on aid.

Power and borders

41 A state is the area of land, of an independent country, with well-defined boundaries, within which there is a politically organised body of people under a single government. A nation is large group of people with strong bonds of identity, united by shared descent, history, traditions, culture and language.

42 An inverse relationship: the higher the index of state fragility the lower the level of development.

43 'Sovereignty' is the absolute authority which independent states exercise in the government of the land and peoples in their territory. 'Territorial integrity' is the principle that the defined territory of a state, over which it has exclusive and legitimate control, is inviolable.

44 'Norms' are the moral principles, customs and ways of living that are universally accepted as standard behaviour. These include the internationally accepted behaviour of state governments, such as their responsibility for maintaining the global system of sovereign states with bounded territories, and protecting their citizens.

45 The resilience of a state is its capacity to resist, absorb and recover from the effects of violent conflict or political upheaval. This depends on the relative fragility or strength of its state apparatus.

46 The challenges include contested territory, current political boundaries in relation to demands for separatism, factional or sectarian tensions, transnational movement of terrorist and extremist activity, the legacy of colonialism in ethnic partitioning, the negative impacts of TNCs, and the binding requirements of international and regional laws of supra-national organisations such as the EU.

47 'Global governance' is the intervention by the global community, in all its formats, attempting to regulate issues, such as human rights, sovereignty and territorial integrity.

48 The international community includes all countries whose identity and sovereignty are recognised by the UN, plus other international organisations, that choose to participate in global discussions and decision making and who act collectively to resolve humanitarian issues.

49 UN involvement includes its peacekeeping missions in areas of conflict over sovereignty and territorial integrity issues, coordination of the work of other international/regional organisations including NGOs, the establishment and reinforcement of norms, treaties and international laws, and resilience building in terms of state apparatus, human rights and local communities.

50 NGOs often work in conflict zones where sovereignty and territorial integrity have been challenged, to assist local communities and governments. Their role, in cooperation with other organisations, is significant in monitoring violence, education regarding norms and laws, training in agricultural techniques, medical assistance, and reinforcing democratic processes.

51 Long-term benefits of intervention include restoring territory according to international law, supporting transition to democracy, improving the business environment and trade relations, establishing sustainable food and water supplies, and re-establishing state authority and all elements of the state apparatus.

52 Military intervention can have unintended effects such as causing further civilian casualties, population displacement, damage to property and infrastructure, food insecurity, disruption of children's education, dependence on aid, and escalation of violence and human rights violations.

Index

Note: **bold** page numbers indicate defined terms

Index